Robert Henry Thurston

Stationary steam engines

Especially as adapted to electric lighting purposes

Robert Henry Thurston

Stationary steam engines
Especially as adapted to electric lighting purposes

ISBN/EAN: 9783337269586

Printed in Europe, USA, Canada, Australia, Japan

Cover: Foto ©Andreas Hilbeck / pixelio.de

More available books at **www.hansebooks.com**

STEAM ENGINES;

ESPECIALLY AS ADAPTED TO

ELECTRIC LIGHTING PURPOSES.

By ROBERT H. THURSTON, A. M., C. E.,

FORMERLY OF THE U. S. N. ENGINEER CORPS ; PROFESSOR OF ENGINEERING AT THE STEVENS INSTITUTE OF TECHNOLOGY; PAST PRESIDENT AMERICAN SOCIETY MECHANICAL ENGINEERS; MEMBER AMERICAN SOCIETY CIVIL ENGINEERS; AMERICAN INSTITUTE MINING ENGINEERS, ETC.

New York:
JOHN WILEY & SONS,
15 ASTOR PLACE.
1884.

PREFACE.

THE following pages were written at the request of the editor of the *Electrician and Electrical Engineer*, from the columns of which periodical they have been reprinted. It was desired that a concise, yet tolerably complete, account should be given of the changes which have taken place during the years which have elapsed since the modern steam engine began to find application in modern industries, and especially those which have been brought about by the demands of the last established of those industries—electric lighting—and which have been the result of unprecedentedly exacting conditions as to efficiency, regulation, and smooth action at high speed of rotation.

The subject was deemed to be of great importance to all mechanical engineers in any way connected with this latest department of their professional work ; and it was desired that a systematic outline should be given of the history of the development of the now standard type of engine for "electric lighting plants" from the older, and still standard, types of stationary engine, in such form that readers should be able to compare the causes of such changes, with the modifications of construction and operation to which they have given rise. This the writer has endeavored to do, giving a philosophical account of this evolution of the later forms of engine, and of the stage to which it has progressed, while, at the same time, presenting correct descriptions of the forms and construction of representative machines. It has been his endeavor to do full justice to the several engines described, and he has carefully avoided the expression of merely personal opinion in regard to debatable points.

R. H. THURSTON.

STEVENS INSTITUTE OF TECHNOLOGY,
Hoboken, N. J.,
May, 1884.

CONTENTS.

STEAM ENGINES
FOR ELECTRIC LIGHTING PLANTS.

BY ROBERT H. THURSTON.

I.

Historical—The Development of the Steam Engine.

THE growth of the steam engine into the forms now familiar to everyone who takes the slightest interest in this most important of modern mechanisms, has occurred by a series of transitions which is easily traced, and which is especially interesting to every thoughtful mechanic as representing the steps in a steady progression, toward ideal perfection, of which the end is not yet seen.

A century ago, James Watt had just begun to introduce the first engines belonging to a, then, new type.[1] A century before (1698), the ingenuity and practical skill of Captain Savery, had conferred an enormous benefit upon the mining industries, and through them upon the world, by applying the "fire engine" of the Marquis of Worcester to raising water from the then rapidly deepening mines.[2] Savery used steam of 8 to 10 atmospheres (120 to 150 pounds) total pressure, in some cases, and he is entitled to fame as the first to introduce that now familiar concomitant of civilization, the

---

1. History of the Growth of the Steam Engine. International Series. N. Y., D. Appleton & Co.
2. The writer finds this engine described in Harris's "*Lexicon Technicum*," of 1704, of which, now rare work, he is so fortunate as to possess a copy.

steam boiler explosion. The usual pressure was 3 atmospheres. These engines demanded about 30 pounds of coal,
per horse-power per hour, as a minimum. The apparatus
of Savery was not what would to-day be called a ·steam
engine, at all. It was not a train of mechanism, involving
moving parts, cylinder, piston, crank and fly-wheel, but
either a single pair of closed vessels, or three vessels, one of
which was a boiler, and the other, or others, metal chambers
of spherical, cylindrical, or ellipsoidal form, which were at
once condensers and pumps. The latter were filled with
steam, which being condensed, the water rose into, and
filled them, and was then forced out by a succeeding charge
of steam, of pressure exceeding that of the head against
which the lift took place. Huyghens (1680), and Papin
(1690), proposed true engines with steam pistons traversing
their cylinders, and forming, on the whole, much such a
train of mechanism as is now so well known[3]; but the
Newcomen engine was the first of this type to come into
practical use. This machine, then called the " Atmospheric
Steam Engine," consisted of a steam cylinder, with a piston
taking steam beneath, the upper end of the cylinder being
open to the atmosphere, the piston actuating a " working
beam," or " walking beam," and, through the latter, working pumps attached to the opposite end. Neither crank,
shaft, nor fly-wheel was used ; the action of the engine was
controlled entirely by the adjustment of its valves. In its
operation, steam at a little higher than atmospheric pressure,
was admitted below the piston ; the weight at the pump end
depressed that extremity of the beam, raising the piston.

3. Mem. Acad. Sci. Paris, 1680. Acta Eruditorum. Leipsic, 1690.

The steam below the piston was then condensed by a jet of water thrown into the cylinder, producing a vacuum; and atmospheric pressure finally forced the piston down, raising the pump-rod and plungers. The weight on the latter was adjusted to the work, so that, when steam was admitted, this weight should force the pumps to discharge the water. The only function of the steam was the displacement of the atmosphere, or counterbalancing it, by entering below the piston, and thus permitting the formation of a vacuum. A writer of that time states[4] that "Mr. Newcomen's invention of the fire engine, enabled us to sink our mines to twice the depth we could formerly do, by any other machinery"; but "every fire engine of magnitude consumes £3,000 worth of coal per annum." The coal consumption was, at best, about 20 pounds per hour and per horse-power. Smeaton, the greatest civil engineer of his time, put up many of these engines in Holland and elsewhere, as well as in Great Britain ; some were 66 inches in diameter of cylinder, and 8 to 9 feet stroke of piston. It was this engine that Watt found in operation, when he entered upon the stage.

Watt was not simply a mechanic ; he was a real philosopher, and a truly scientific investigator. A model Newcomen engine, having been brought to him to be repaired, he took advantage of the opportunity to study the principles of its construction, to ascertain its defects, and to devise proper remedies. He found that the sources of loss were the conductivity, and radiating power of the steam cylinder, the alternate heating and cooling of the metal at each stroke, the imperfect vacuum, and the wastes from boiler and

---

4. *Mineralogia Cornubiensis.* Price. 1778. Appendix.

steam pipes. To correct these defects, he clothed his boilers and steam pipes with non-conductors, sometimes even making boiler shells of wood. Smeaton had already covered the pistons and cylinder heads with wood. Watt made a small wooden steam cylinder, and obtained great economy ; he made a more practicable improvement, however, when he devised the steam jacket. He attached a separate condenser to prevent the loss due to the introduction of condensing water into the steam cylinder, closed the cylinder at the top, made the engine double-acting, and finally adapted the engine to drive machinery, fitting it with shaft and fly-wheel, throttle valve, and governor, and thus making the steam engine such as we see it to-day, in all essential particulars, not excepting the steam jacket, and the arrangement of its valve gear to secure economy by the expansion of the steam.   His engine was substantially complete by the year 1784.[5]

Later changes have been a succession of refinements, and of developments in application.   Stephenson, and his contemporaries, applied steam on railroads ; Stevens, Fitch, and Evans, and, finally, Fulton, in the United States, and Bell and others, in Europe, introduced steam navigation ; Sickels invented the "detachable" cut-off valve gear ; Corliss introduced the peculiar type of engine that has given him a world-wide fame, and so attached its governor as to determine the point of cut-off automatically, and thus to regulate the engine ; and, a little earlier, Robert L. and Francis B. Stevens designed the American river steamboat, and its beam engine, with so simple and effective a valve

5. History of the Growth of the Steam Engine. P. 119. Farey on St. Engine.

gear that it remains, to-day, still standard.   The compound
engine, even, was brought out by contemporaries of Watt,
and thus every prominent feature and essential detail of
the modern steam engine was introduced at, or before, the
middle of the nineteenth century.

Yet, practice has been steadily changing during the century, and the form and proportions of the steam engine, and
the methods of steam distribution, have been undergoing
constant changes.   In the time of Watt, steam was worked
at about 7 pounds pressure, per square inch, in stationary
engines; they were always fitted with condenser and air-
pump, and were slow in movement, and were, consequently,
of small power in proportion to their size; they wasted heat
and fuel to such an extent, as to demand 6 or 8 pounds of
coal per horse-power and per hour.   It is true that Wolff,
in 1804, expanded 6 or 8 times, using higher steam and obtained the horse-power with 4 pounds of fuel per hour, and
that John Stevens and Oliver Evans, in the United States,
and Trevithick, in Great Britain, had already used still
higher steam in non-condensing engines; but these examples
simply illustrated the fact, now familiar to every student of
philosophical history, as pictured by Draper, Buckle and
Whewell, that isolated examples which lead standard practice
by a half century or more, are to be observed during the
growth of every art.   Recognized standard practice is always as conservative as it is permitted to be by trade competition, and usually changes very slowly.   Principles may
be discovered and understood, and a correct theory of design and of practice may be made generally familiar, and
often is, in a brief period; but the growth of application

and the familiarizing of constructors and operatives with new mechanisms, and new methods of management, requires time, and is slow at best. Thus it has happened, that although the principles of steam engine economy were, in the main, well understood by James Watt, and some of his competitors, nearly a century ago, and have become well settled in later years, we are still far from a completely satisfactory solution of the problem, which, as stated by the writer elsewhere, may be enunciated thus:—' To construct a machine which shall, in the most perfect manner possible, convert the energy of heat into mechanical power, the heat being derived from the combustion of fuel, and steam being the receiver and conveyer of that heat."

II.

Principles of Economy; Special Requirements.

THE principles of economical working, noted by James Watt, and plainly stated by him, were but slowly recognized by others, and the improvement of the steam engine was, for many years, correspondingly slow. The principles that must govern the engineer, in the attempt to secure highest efficiency, may be summarized thus:

1. The greatest practicable range of commercially economical expansive working of steam must be adopted; the fluid must enter the cylinder at the highest admissible pressure, and must be expanded down to the minimum economical pressure at exhaust.

2. The wastes of heat must be made the least possible; all loss of heat by conduction and radiation from the engine must be prevented, if possible, and the usually much more serious waste which occurs within the engine, by transfer of heat from the steam side to the exhaust, by "cylinder condensation" and re-evaporation, without doing its proportion of work, must be checked as completely as is practicable. This latter condition, as well as commercial considerations, limits the degree of expansion allowable. It also dictates high speed of engine.

3. The largest amount of work must be done by the engine that it can perform, with due regard to the preceding conditions. This condition compels us to drive the engine up to the highest safe speed, and to adopt the highest practicable mean steam pressure.

The first two of the above requirements give maximum efficiency of fluid, consistent with commercial economy, and the latter gives highest efficiency of machine. In addition to these requisites, which are not peculiar to any style of engine, or to any one of the innumerable applications of steam power, the adaptation of the machine to driving the dynamo-electric apparatus of an electric lighting plant, compels the designing and constructing engineer to meet certain demands which, although not peculiar to this work, are, nevertheless, more imperative here than elsewhere. The principal of these requirements are effective regulation, compactness, simplicity of parts, strength and durability, and small cost, both of original purchase and of repairs. In the attempt to meet these demands, the modern "high speed engine" has gradually taken shape.

In the time of Watt, a pressure of seven pounds of steam, with condensation, and a low piston speed, equal, usually, in feet per minute, to about one hundred and twenty-eight times the cube root of the length of stroke, according to Watt's own rule, represented standard practice. As time went on, steam pressures and piston speeds gradually rose, and when, in 1849, Corliss brought out the typical modern "*Drop Cut-off Engine*," pressures of sixty pounds, and speeds of piston reaching 450 feet per minute were becoming usual. At such speeds, the "drop cut-off" was thoroughly effective, and the steam valve, detached from the driving mechanism, fell into its seat with sufficient promptness and accuracy, as to time of closing, to do good work; the governor had no other work to do than to detach the valve, and was thus able to regulate with an exactness that is still beyond competition. These engines are very extensively used to drive the smaller electric light machines, and particularly where a considerable number are to be driven together; they are not adapted to the work of driving the large "dynamo," where it is desired to couple direct from crank-shaft to armature.

As piston speeds increased, the drop cut-off became less satisfactory, where the load was variable. It became slowly understood, among builders and users of engines, that one important element of economy of fuel and cheapness in cost of engine is the maximum speed of engine consistent with endurance and safety. Speeds were, after a time, rapidly increased, the Porter-Allen engine leading in this movement, and small engines, working at high speed, displaced large engines of the older type. It soon became evident that this change must lead to the re-introduction of the "positive motion" classes of valve gear and expansion gear that Sickles, Corliss and Green had temporarily displaced, notwithstanding the fact that these builders had greatly increased the speeds of their engines. All the so-called "high-speed engines," which are best known in the market, are of this later type. The slower running engines are nearly all fitted with governors of the fly-ball class, geared, or belted, to revolve at a much higher speed than the engine itself ; but the great velocity of rotation of the new engines, from 200 to 500 revolutions per minute, in the small sizes, and often a piston speed of about 800 times the cube root of stroke, permits the attachment of the governor directly to the shaft; and this is done in the later styles. This change of position of the governor, in turn, has led to a change in its construction. The balls, instead of being hung from a vertical revolving spindle by arms pivoted on that spindle, are attached to arms carried on the main shaft, or the driving pulley, and revolve in a vertical plane at right angles to the shaft; they are held in place against the action of centrifugal force by springs, and arranged to adjust the eccentric, and to vary the expansion, in a manner which will be plainly seen when studying their construction in the later sections of this paper, in which these engines will be described with the aid of carefully made engravings. The high speed engine, as adapted to the work of directly driving the "dynamo," therefore, may be described as a high pressure, non-condensing engine, of short stroke, and high speed of rotation, with a positive-motion valve-gear, and regulated by a governor, which is usually mounted on the shaft, and so attached as to alter the expansion by varying the lead of the valve. Its essential features are high speed of rotation, good regulation by a positive gear, economy, simplicity, and compactness. It is this engine only, which is found to do good work under these peculiarly exacting conditions.

It is proposed to study the best known engines of this and the earlier classes, and to compare them, with a view to bringing out their peculiarities and their special merits, while the purchaser will, besides, study the machine which he proposes to buy, to determine whether its material and workmanship are as excellent as are the principles of its design.

The conditions demanded can here be merely outlined, in the following resumé[1] of the requisites of successful practice:

---

1. Report on Machinery and Manufactures. R. H. Thurston. Vol. iii. Reports of the Scientific Commissioners of the United States to Vienna; 1873.

1. A good design, by which is meant:

    a. Correct proportions, both in general dimensions and arrangements of parts, and proper forms and sizes of details to withstand safely the forces which may be expected to come upon them.   •

    b. A general plan which embodies the recognized practice of good engineering.

    c. Adaptation to the specific work to be performed, in size and in efficiency. It sometimes happens that good practice dictates the use of a comparatively un-economical design.

2. Good construction, by which is meant:

    a. The use of good material.

    b. Accurate workmanship.

    c. Skillful fitting and a proper "assemblage" of parts.

3. Proper connection with its work, that it may do that work under the conditions assumed in its design.

4. Skillful management.

In the endeavor to secure these requisites, it is generally advisable to use steam at a pressure not far from one hundred pounds per square inch. The benefits of increasing pressure diminish so rapidly above this point, that it is not yet certain whether it will, with existing engines, pay to carry pressure much higher. The ratio of expansion is to be determined with reference to this pressure, as well as to size of engine. It will usually be found even more wasteful to cut-off too short than to "follow" too far; and Rankine's principle of adjusting this point by consideration of the relative cost of large and small engines, as well as the principles controlling the economy of fuel, dictate, that for these engines, which are nearly always non-condensing and un-jacketed, the ratio of expansion must usually be low—say from three to five, as higher pressures range from sixty to one hundred pounds per square inch[2]—and that the terminal pressure shall usually be kept some five or ten pounds above that of the atmosphere.

Moderate "superheating" is found advantageous ; but it is seldom carried beyond about a hundred degrees above the normal temperature of the steam.   " Steam jacketing," as practiced in nearly all compound engines, is of advantage; but is not usually considered to pay for the added cost and risk in engines of the class here considered, and especially in high-speed engines.   The "compound" engine has not found a place in this field.   Smeaton's idea—or rather Watt's, first attempted on a large scale by Smeaton—of surrounding the working fluid with non-conducting surfaces, is not yet found practicable with the high steam pressures and temperatures now usual.   Its final adoption, however, is beyond doubt, as it is a far more promising system of economizing heat, now wasted, than either superheating or steam jacketing.   The latter, indeed, is a method of introducing a waste to check greater loss.

Careful protection of external heated surfaces of the cylinder against losses by conduction or radiation, is always practiced where it can be conveniently done, and parts which cannot well be so covered are highly polished.   A well polished surface transmits very little heat.

Back pressure, a frequent cause of waste of power, is

2. Ibid. Pp. 17, 18, 49.

reduced by making the exhaust parts large, and the exhaust opening of the valve rapid, and by giving "lead" to the exhaust, so that the steam shall leave the cylinder just before, rather than just after, the return stroke begins.

Friction is reduced to a minimum by carefully proportioning the journals, and by securing free and continuous lubrication with a good oil or grease.

An engine in which all the above requirements are fully met is certain to be a good machine.

It is not proposed to compare the steam engine with the gas engine, or with other motors. The gas engine is, in many cases, likely to prove useful in consequence of its compactness, cheapness of first cost, freedom from risk, and small expense for attendance ; but it is expensive in use of fuel, and is rarely as little liable to annoying interruptions of operation as the steam engine, and also possesses other minor disadvantages. Nevertheless, Otto and Clerk, and other inventors and constructors, have greatly improved this machine of late, and have brought the expenditure, in ten-horse engines, down to twenty cubic feet of gas, or less, per hour and per horse-power ; and although this is still double the theoretical figure, no one can say how soon the latter consumption may not be much more closely approximated to. The gas engine is certain to find work in this direction. Hot air engines, as yet, give less promise ; but it would be rash to predict their total exclusion from the field.

Water-wheels, especially when used exclusively for supplying power to the lighting plant, are, where available, thoroughly satisfactory prime movers.

In studying the steam engine from the standpoint here

taken, we will divide them, first, with reference to their method of driving the dynamo-electric machine, into two classes :

1. Engines which may be used in driving by belt, and which are not adapted for direct connection.

2. Engines especially designed and constructed to be coupled directly to the " dynamo."

The first class of engines is in very extensive use, and is, by many of the more conservative engineers, still preferred to the second. The latter constitute the so-called "modern " type of engine, and are gradually coming into use, some engineers adopting them, both for direct and for indirect connection. The best engineers are not yet fully in accord in regard to the question, whether they have passed the experimental stage.

III.

Engines Indirectly Connected, only.

THE CORLISS ENGINE.

DIVIDING engines used in driving dynamo-electric machines into two principal classes—engines driving indirectly through gearing or belting, and engines directly connected to the armatures—we may profitably devote considerable space to the first class. And, although machines of the kind which have come to be distinguished by the appellation "high-speed engines" may be, and often are, indirectly connected, it is proposed to leave the examination of such engines to a later article on directly connected engines, and here to describe only the "drop-cut-off" engines, or those with "detachable valve-gear," which can only drive the armature of the "dynamo" indirectly.

The first drop cut-off introduced, had a form patented by Fred. E. Sickles, in 1841. This engine was first built for mill purposes, by Thurston, Gardner & Co., at Providence, R. I., that firm then holding the Sickles' patents, except that the marine engine business was retained by Sickles. The modern stationary engine was thus introduced, and was soon extensively made known among steam users by its superior performance when competing with the older engines, which were then usually arranged to expand steam about one and a half times by the lap of the single three-ported valve. A few engines were built of a better design, fitted with an independent cut-off valve on the back of the main

valve. These two last named engines would, at best, with
good boilers use five or six pounds of coal per hour, and
per horse-power, where the Sickles valve-gear would bring
the consumption down to four.

Regulation was always effected by a governor controlling
a throttle valve. This governor was usually a common fly-
ball governor, and its deficiency in power and lack of
isochronism, the distance of the regulating valve from the
engine valves, and the range of motion required in its
operation, and the resistance offered by the packing of the
steam, altogether, made this combination a very ineffective
regulating apparatus. Thurston, Gardner & Co., sub-
stituted for this the Pitcher hydraulic regulator and a
register valve, which gave a much better regulation; this
contrivance was also isochronous, *i. e.*, it was capable of
holding the engine at speed, whatever the variation of
steam-pressure or of load.

But an immense step in advance of this, then, best practice was made by Geo. H. Corliss, a young Providence mechanic, who had exchanged the *rôle* of sewing-machine inventor for that of the inventor of the most famous steam engine
that has appeared since the time of Watt. The Corliss engine
was patented in 1849, and rapidly came into use, its remarkable economy, when competing with the best existing
engines, the peculiar business tactics of its builder, and the
rapidly increasing demand for efficient, and especially well
regulated, engines, combining to give it a wonderfully rapid
introduction.

The engine is an interesting illustration of a machine
which is the representative of a peculiar type, each detail

of which is especially adapted to its place in that machine, and is characteristically different from the parts which perform the same office in other engines. The leading features of this machine are:

1. The use of four valves—two steam, and two exhaust —so placed as to reduce "clearance" to a minimum.

2. The use of a rotating valve, capable of being cheaply and readily fitted up, of being easily moved, and of being conveniently worked by connections outside the steam spaces.

3. The use of a "wrist-plate," caused to oscillate by a single eccentric, and directly so connected with all four valves that each may be given a rapid opening and closing movement, and be held open and nearly still, at either end of its range, by swinging the line of connection nearly into the line between centres, thus permitting nearly a full opening of port to be maintained during an appreciable interval, and a free and complete steam supply and exhaust.

4. A beautifully simple and effective method of detaching the steam valve from the driving mechanism, and of insuring its rapid and certain closure at the proper moment, to produce any desired expansion of steam.

5. A direct connection of the governor, so as to determine the ratio of expansion, while so adjusting the power of the engine to the work to be done, that the variation of speed with changing loads becomes a minimum.

6. Making this latter adjustment in such a way as to throw the least possible work on the regulating mechanism, and thus to give the governor the greatest possible sensitiveness and accuracy of action.

7. A form of frame and general design of engine, which gives maximum strength and stiffness, with least cost and weight.

All these features are combined to form a steam engine essentially different, in general and in detail, from the engines contemporary with or succeeding it, except where the latter may properly be classed as Corliss engines. It rarely happens that an inventor succeeds in originating a plan so wholly and so essentially novel; and it is still less frequently the fact, that a peculiarly original device is found superior to all competing machines. In operation, the engine was found to exhibit a remarkable economy of fuel. and a singularly perfect regulation, and to be far more durable and more economical in cost of repairs, on the average, than rival builders supposed possible. It very soon took the leading place in the market.

The inventor established himself at Providence, and put in operation a method of marketing his machine which was as novel and as successful as the mechanical device itself. He offered to put his engine in place of rival engines, either with a guarantee of a certain saving, and at a stipulated price, or, often, to take as his compensation the actual saving shown on the books in a stated time. This system was eminently satisfactory to the purchaser, both as making him safe against loss, and as giving him some of that confidence in the engine which the maker himself unquestionably possessed. Corliss' work fully justified his claims, and the expenditure of fuel was brought down to between three and four pounds per hour, and per horse-power, according to size and situation of the engine, with occasionally much better figures in condensing engines. This engine is now built, not only by the Corliss Steam Engine Co., under the eye of the inventor, but by many other builders. It has found its way into every part of the world ; and the engineer visiting Europe will find a pleasure in observing the general adoption of this American invention in every country, and for every purpose. European makers frequently modify the design, but rarely with the desired effect of securing an improvement in cost or efficiency, and very often with a decidedly contrary result.

Corliss engines are now very frequently adopted in electric lighting, and are always belted to the dynamos. Their excellent regulation is as important a feature in this application, as is their economy in use of steam. When carelessly constructed, they are, of course, likely to prove wasteful and irregular in action. But that these engines can be made to give very perfect uniformity of rotation will be evident, when it is stated that the writer, in testing engines of this class, has found that the variation of speed was so slight as to be practically inappreciable, even when the amount of work thrown on or off, was a very large proportion of that done by the engine when working at its rated power.

One other reason for the success of this engine is unquestionably the comparatively small cost of its construction, where competing with the earlier forms of engine with detachable valve-gear. Its valve-faces, particularly, and their seats, are surfaces of revolution, and they, as well as a large part of the finished work about the engine, being almost wholly lathe-work, the cost of fitting up is comparatively small.

In detail, the engine consists, as shown in the illustration, page 19, of one of its standard forms, of a steam-cylinder sustained by any substantial connection with the foundation. The main pillar-block sustains the crank-shaft at the opposite end of the machine, and a strong brace, connecting these two pieces, forms, at the same time, a support for the crosshead guides.

The four valves are placed at top and bottom of each end of the cylinder, their rotating stems projecting, and are moved by the "wrist-plate," set usually, as here, at the middle of the cylinder, the valve connections radiating to the four corners, where each is attached to the valve rocking-arm, the exhaust by pin-connections, the steam by a catch, which can be readily " tripped " by the adjustment of a little cam set on the valve-stem, behind the arm. When tripped, the steam valves are closed by a spring, or in engines now built by Mr. Corliss, by a "vacuum-pot," and by weights in his earlier engines, and in those of other builders.

The governor is belted from a pulley on the main-shaft, and its oscillations are controlled by a "dash-pot," seen attached to the side of its standard. The governor, having no work to do but to set the tripping-cam, or the equivalent for it adopted by Corliss and others in various designs, is entirely free to adjust itself to the normal position due the speed of the engine, and thus is made perfectly capable of doing the best possible work. Many foreign builders have attached the Porter loaded governor to this engine. The advantage is less obvious here than in engines in which more strength of action is needed.

From what has been stated, it is seen that the Corliss engine came into use in consequence of its combination, to an extent up to that time unequalled, of several special features. Some of these points are not necessarily peculiar to the Corliss type of engine; but they, nevertheless, were peculiar to that engine at the time of its introduction. The main points were : the rapid and wide opening of the steam and exhaust openings; the shortness and directness of the ports; the resulting small clearance and "dead" spaces ; the quickness of closure of the steam valves ; the adaptation of the main valve to the functions of a cut-off valve ; the connection of the governor to the cut-off gear in such a manner as to determine the point of cut-off without being itself hampered by the connection ; the location of the exhaust ports at the under side of the cylinder so as to drain the cylinder thoroughly ; and the simple, easily constructed form of the machine and of its details.

The general form of the engine has been preserved by nearly all copyists, and the parts of the valve gear and details of regulating mechanism have been seldom much modified. A few builders have, however, made changes which are worthy of notice, but which we have not time or space to study as they deserve.

The action of the Corliss engine is as follows :[1]

The valves are driven by the eccentric rod through the "wrist-plate," *E*, vibrating on a pin projecting from the

1. History of the growth of the Steam Engine. D. Appleton & Co., N. Y. 1878.

cylinder. Links, $E\,D$, $E\,D$, $E\,F$, $E\,F$, take motion, from properly set pins on this wrist-plate, to the steam valve rock-shafts, D, D, and to the exhaust valves, F, F, moving them with a peculiar varying motion in such a manner as to open and close the ports rapidly, and to hold them open, when the valves are off the ports, in such a way as to give the least possible loss of pressure during the exit or the entrance of steam. The links leading to the steam valves are fitted

THE CORLISS ENGINE.

with catches, or latches, which may be disengaged, as the valve opens, at any desired point within about half stroke; and the time of this disengagement is determined by the rotation of a cam seen on the valve stem above D, which cam is rotated by the governor through the rod H, leading off to the left. The slowing of the engine, in consequence of reduced steam pressure or of increased load, causes the catch to hold its contact longer and the steam to follow

farther, and the reverse.   When the catch is disengaged,
the valve is closed by a spring or weight attached to the

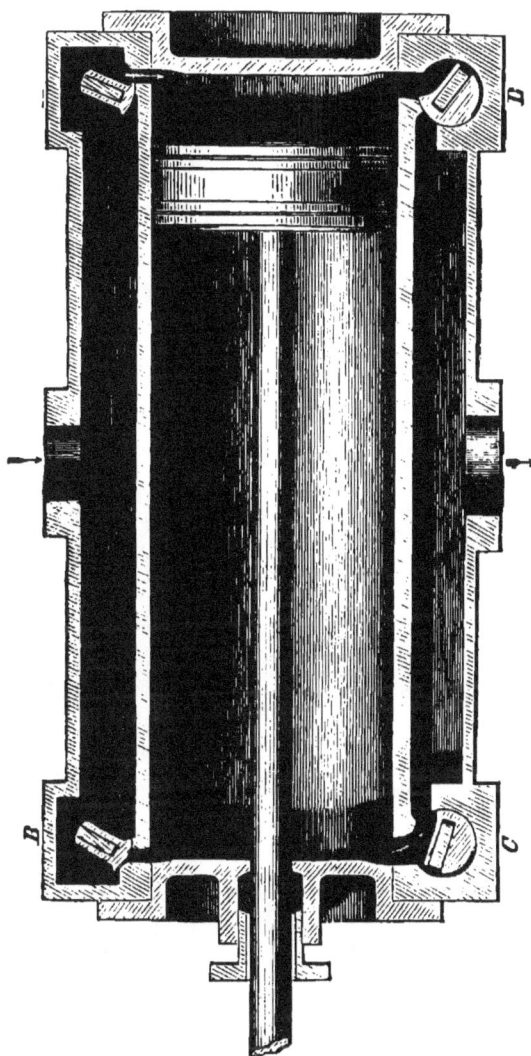

THE CORLISS ENGINE CYLINDER.

vertical rods seen connected to the rock-shaft arm.   Corliss
uses a device in place of this which is not here shown.   The

dash-pots are under the floor, in the case here illustrated, or on the column supporting the governor in the engines just referred to.   It is always an air dash-pot.   The device invented by Sickles was a water dash-pot.

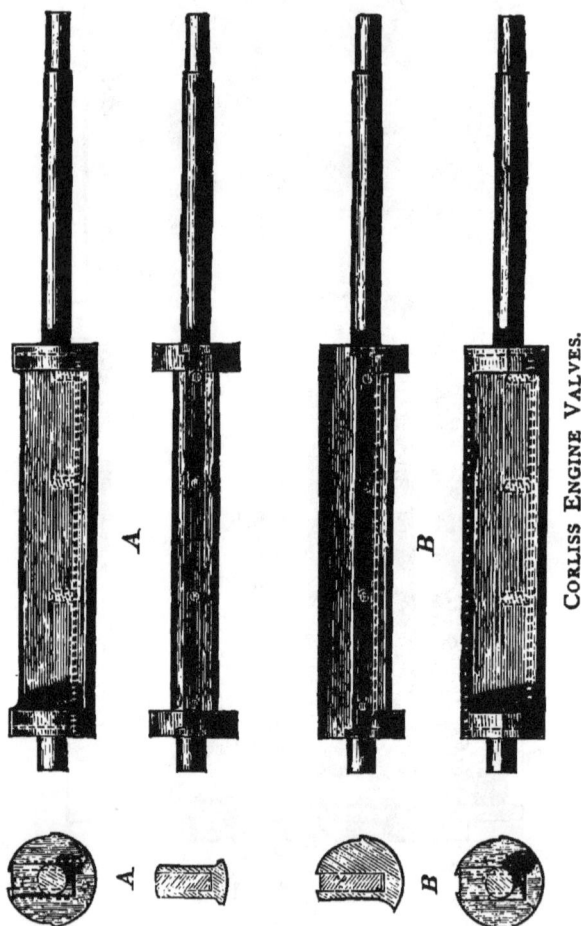

CORLISS ENGINE VALVES.

The standard form of Corliss valve is very well exhibited by the illustrations here given, which are taken from the drawings of Mr. Harris.

Those marked *A* are the steam, and those marked *B* are the exhaust valves. Both consist, as is seen, of cylinders, parts of which have been cut away, leaving the working and bearing surfaces of no greater extent than is necessary to subserve the purposes of the valve. These surfaces are of the simplest possible form and are easily fitted up in the lathe. In order that they may come to a bearing with certainty, and without regard to the position of the spindle relatively to the valve, they are made with a longitudinal slit into which fits, without jamming, the blade of the rock-shaft. The valves are thus allowed to come to a bearing, and even to wear down in their seats without causing leakage.

The next Fig. shows the arrangement of this valve as seen in longitudinal section of the chest. As this maker

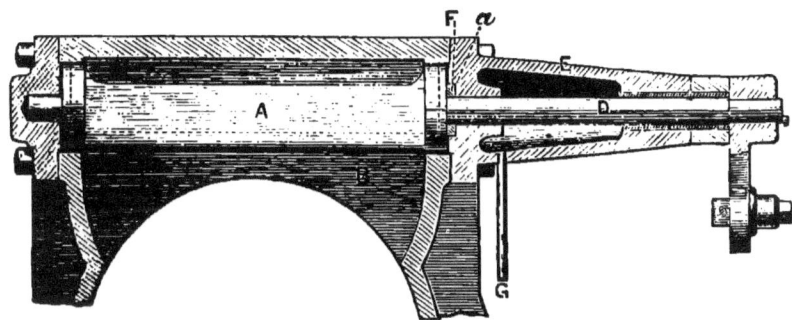

HARRIS-CORLISS VALVE.

constructs it, the stem goes through a fitted opening, without stuffing box, and the slight drip is carried off from the closed space at *D*; thus none escapes into the engine room. The steel collar at *F*, which is shrunk on the stem, fits into the recess at *a* and serves as a packing. As the tendency of the stem to shift outward always causes the collar to wear to a fit, it is not likely often to wear leaky. Another detail of interest in the Corliss engine is the

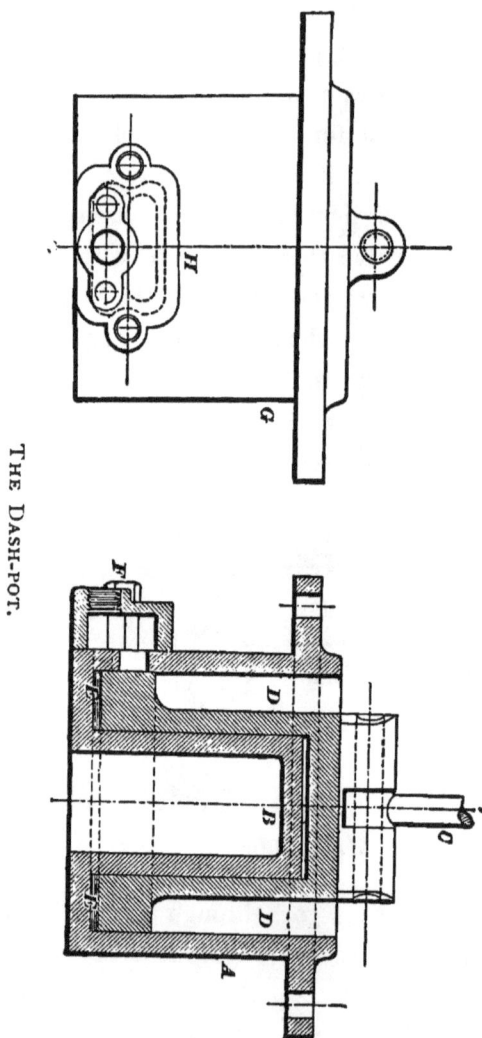

THE DASH-POT.

"dash-pot." When the valve is suddenly closed, some device is necessary to prevent jar at the instant of its coming to rest. This device is the dash-pot. The form adopted by Corliss consists of a shallow cup into which a piston on the valve stem fits, cushioning the enclosed air, and thus checking the motion of the valve without shock. This dash-pot, made by Watts, Campbell & Co., who have successfully introduced Corliss engines into electric light establishments in New York city and elsewhere, is that seen in the Figs.

The annular piston, E, E, fits the cylinder, D, D, E, E, and a space, seen above B, forms a vacuum chamber which assists the spring or weight, closing the valve by the formation of a more or less complete vacuum, as the piston is raised while the valve is opening. A small cock, not seen, is arranged to adjust the degree of exhaustion of this chamber. When the valve has nearly reached its seat, the piston D, passes the opening from F into the outer space and the enclosed air then acts as a cushion, checking the movement of the valve. In the engines of these builders, great care is taken to keep the cold exhaust steam clear from the cylinder as it passes out, in order to prevent the condensation which occurs where this precaution is neglected.

Many Corliss engines are already at work driving electric lighting apparatus, and are giving good satisfaction, according to the testimony given the writer by the officers of the companies using them. One, built by the Corliss Steam Engine Co., is at work at Providence, R. I., driving nine dynamos, and a number are in use in New York city, and other large cities of the United States and of Europe.

At how high a speed they can be operated with satisfaction to the user is not definitely known. The writer has known one of these engines, coupled to a fast running ro'l-train, to be driven without apparent difficulty for several years at a speed of 120 revolutions per minute, although of four feet stroke.   This engine is still running.   Those who use, as well as the engineers who build, this class of engines, however, are apt to be conservative and to prefer the moderate speeds with indefinite endurance, to higher speeds with a shorter life of engine and greater cost in keeping in repair; and to consider that the satisfaction of having a prime motor, which is not likely during their business lives to give them any trouble, is more than a compensation for any possible saving in dollars and cents to be effected by the adoption of the higher velocities of piston and of crank-shaft rotation.

THE WHEELOCK ENGINE

is an ingeniously arranged engine of the class considered in this division of the subject.

Its form is seen in the accompanying engravings.

The steam chest is placed below the cylinder and the steam and exhaust valves are set side by side, the latter serving both as induction and eduction valve, and having the same action, nearly, as the common three ported slide valve, while the function of the former is principally that of a cut-off valve.   The latter, or main valve, is set nearest the end of the cylinder and the exhaust steam is thus permitted to escape directly and promptly from the engine.   The valves are coned, slightly, and may be adjusted to take up wear, or to relieve pressure on their seats.   These valves

THE WHEELOCK ENGINE.

are carried on steel trunnions, and with hardened surfaces of contact are but little subject to wear. The steam or cut-off valve is set further away from the cylinder than in the standard arrangements of Corliss and other builders of that class of engines, and this enables the maker of this engine to secure a single port with reduced clearance and less liability to leakage, should the expansion valve leak. In this engine—and it should be the case in every engine in which the regulator is driven by belt—the connection from shaft to governor is so made that the breaking of the belt permits an automatic closing of the valve and the stopping

THE WHEELOCK VALVES.

of the engine. The regularity of motion of the class of engines described in this section, may be inferred from the fact stated in regard to the engine here studied, that it has been known to vary but a half revolution per minute when five-sixths of the load was thrown off.

Engines of the class described in this section have displayed an economy in the use of fuel that has been rarely equalled by the best type of compound engine, working under the same conditions of steam supply. With good boilers, they have given the horse-power with a consumption of two pounds an hour for condensing engines, and three pounds for non-condensing engines. They have quite often demanded but a ton of coal for 100 barrels of flour ground, in well arranged mills; and one and a quarter tons is a very usual figure. A number of good makers are now building such engines, and the purchaser can readily suit himself if desirous of selecting an engine of any grade, either as to cost or excellence of construction. They are well adapted to driving either large or small electric lighting plants; and, if purchased of a reliable maker, may be confidently expected to give satisfaction.

THE GREENE ENGINE.

NEARLY all "drop cut-off engines" are constructed, like those described in the preceding article, with a single eccentric, which drives both the steam and the exhaust valves. Both sets of valves must, therefore, have the same motion relatively to the piston, except so far as their motion can be modified, as in the Corliss engine, by the method of connection of valve and eccentric. They must stop and start at the same instant, and their motion during their travel must be more or less similar. But such a system is controlled in its action by the necessary motion of the exhaust valve. That valve must be adjusted to open and to close very nearly at the beginning and the end of the return stroke, in order that the exhaust may be prompt and free, and that the compression shall be right. The movement of the gear, on the steam side,

THE GREENE ENGINE.

must thus be also one which shall open the valve to take steam at the commencement of the steam stroke, and, if the valve is not tripped, close the port at the end of that stroke. It is further evident, that if the valve is to be detached by its own motion, it can only be tripped during the forward part of its movement, and that, passing that stage, and commencing to return before the cut-off takes place, the valve must be allowed to remain undetached until the end of stroke, and steam must follow full stroke. An engine thus constructed, and so adjusted to its work as to cut-off at about half stroke, will evidently, if the work or the steam pressure becomes variable, be likely to operate very irregularly, at one time cutting off at a little inside half stroke, and then jumping to full stroke. This variation of steam distribution may thus itself introduce a disturbing element, and the engine may give a very unsatisfactory performance. Such an adjustment of power of engine to the work to be done, does not often take place in engines of the class which is here studied, as the best point of cut-off is usually not far from one-third or one-fourth stroke, and the variation in the load is not often great enough to cause serious difficulty in the manner described above.

One advantage possessed by the arrangement of valve gear, thus subject to criticism, is that, should, as sometimes happens, the valve fail to close, or should it lag behind very greatly, in fast running engines, it is certain that it cannot be left open beyond the end of that stroke, as the returning motion of the valve-gear will bring the latch into gear again, and will insure its closing. Mr. Corliss considers this point of sufficient importance to make it inex-

pedient to drive the steam valves by the method to be
described in this article.　It is undoubtedly an advantage
to be able to secure such an arrangement of valve-gear
that the ratio of expansion may be varied by the governor
from the beginning to the very end of the stroke, so that
the engine may adapt its steam supply to any load that
may be thrown upon it, whatever the extent of that variation may be, and to cut-off at any point from end to end
of the stroke.　This can be done by the adoption of a gear
of the class known, for many years past, from the time of
the earliest steam engines in fact, as the "plug-tree" form of
valve-gear.　It was this class of gear that was used on engines
before the days of Watt, that greatest of inventors, for pumping out the deep mines of Great Britain—the Newcomen
engine.　It may be still seen in use on all so-called Cornish
engines, which are to be found in the water works of this and
other countries—the most costly, cumbersome, and unsatisfactory style of engine which has been applied to that kind of
work in modern times.　The distinguishing feature of this
gear, is, that it is so adjusted, that the motion of the valve is
produced by a mechanism which begins and ends its movement with the action of the piston; in the Cornish engine
it is actuated by the engine beam.　It is easy to obtain a
motion of this character, by the use of an eccentric, by
simply setting it so as to make its throw directly with, or
opposite to, the crank.　In such a case, it is seen that the
exhaust valve must be driven by an independent eccentric,
and the cost of the engine is thus somewhat increased.
This is not a large item, however.　The "Greene engine"
is an engine fitted with such a valve-motion.

In the accompanying illustration,[1] which exhibits this machine, the valves are seen to be four in number, as in the engines already described. They are flat valves, instead of cylindrical, and are thought by the inventor to be better than the latter, as being easier to refit when worn, and as being less liable to become leaky. The cut-off mechanism consists of a sliding bar, *A*, driven by an eccentric, set to

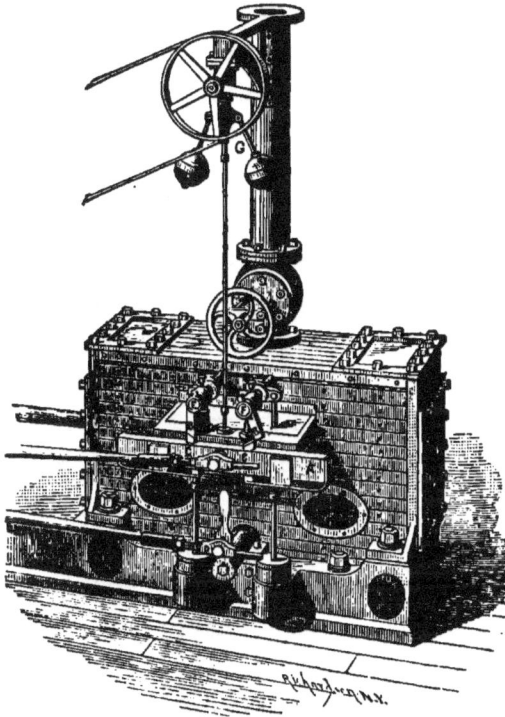

GREENE VALVE MOTION.

give it motion parallel to the centre line of the cylinder, and with a movement co-incident, as to time, with the motion of the piston; of a pair of "tappets," *C, C*, set in this bar

1. Hist. of the Growth of the Steam Engine. D. Appleton & Co. N. Y., 1878.

and adjustable vertically in such a manner as to engage the rock-shaft arms, *B, B,* on the ends of the rock-shafts, *E, F,* which rock-shafts are attached to the valve-links inside the steam chest; of a set of springs which hold these tappets up to their work, and in contact with the "gauge-bar" behind the bar, *A,* and out of sight in the drawing. This gauge-bar is adjusted to the proper height, and is varied in position, as the load varies, by the action of the governor which is connected to the gauge-bar by the rod extending up to it at *G.* The exhaust valves are seen below, and are driven by the second eccentric there shown. They are so placed as to thoroughly drain the cylinder of all water carried into it by priming, or produced by cylinder condensation. The eccentric driving these valves is set at right angles to the position of the crank. In consequence of this independence of the two sets of valves, this engine can cut-off at any point in the stroke during a complete half revolution of the crank. This form of engine was invented by a Providence mechanic, Mr. Noble T. Greene, and was patented in the year 1855. Mr. Greene, then of the firm of Thurston, Greene & Co., introduced this engine a few years after the merits of the drop cut-off had been proven by Sickles and Corliss so fully that it was easy to secure a market for new devices of this class; and the introduction of this engine has had much to do with the rapid progress of these more economical kinds of engine.

The form of the engine has been somewhat modified at various times, although its characteristic features have been carefully preserved. The steam valve, as designed by the writer, who, at the time of its first appearance, had an occasional opportunity to exercise his powers as a designer on this engine, is seen in the next Fig.[1]

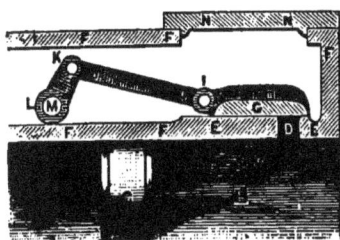

THURSTON'S VALVE.

The valve, *G, H,* covering the steam port, *D,* in the cylinder, *A, B,* is driven by the rod, *J, J,* which is connected to the rock-shaft, *M,* by the arm, *L, K,* in such a manner that the line, *K, I,* will, when prolonged, intersect the valve-face at its middle point *G;* it is thus so set that the line of action of the link, *K, I,* meeting the valve seat directly under the middle of the valve, does not produce any tendency to rock the latter, and thus to cause wear at the edges, or leaks of steam past the valve into the port.

The latest form of the Greene engine, familiar to the writer, is that now constructed by the Providence Steam Engine Co., and shown in the large illustration, page 35. In this engine, the steam valves are connected to the cut-off mechanism, by a set of rods or stems running parallel to their seats, and emerging into the air through stuffing boxes, properly provided with easily set and easy working packing; these valve stems are connected to the rock-shafts, and are driven as in the arrangement already described, very nearly; this design has some advantages over the old, in keeping the working parts, and especially the joints, out of the steam space. The exhaust valves are gridiron slides, set to travel across the line of the cylinder, and driven from

1. Supplied by D. Appleton & Co.

a horizontal rock-shaft, extending forward to the eccentric
on the crank-shaft; the governor is a Porter loaded
governor, driven by a belt from the main shaft; the cut-off
mechanism is illustrated in the last of this series of illustra·
tions.

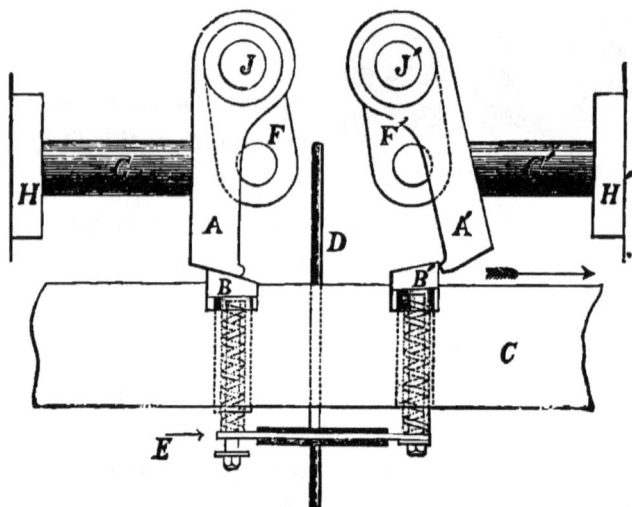

GREENE TRIP MOTION.

The tappets, *A, A*, are carried by the rock-shafts, *J,
J*, which, in turn, drive the arms, *F, F*, and the valves
attached to the stems, *G, G*, passing through the stuffing
boxes, *H, H;* the tappets, *B, B*, engage these rock-levers,
and are adjusted vertically by the governor rod, *D*, and
held up against the gauge bar or the rock-lever, as the case
may be, by the springs set in the sliding bar. When the
speed of the engine is above that for which the engine is
set, the governor, acting through the rod, *D*, depresses the
tappets, and they do not retain their connection with the
rock-lever as long as when at normal speed; when the speed

falls below that fixed by the constructor, the governor rod rises, and the tappets are thus permitted to rise, and to remain in contact with the rock-lever, holding open the steam valve for a longer period than before. The longer the valve is to be kept open, and the farther the steam is to follow, therefore, the wider does the port open to steam. When the tappets travel to the point of cut-off, they swing clear of the rock-levers; the weights, acting together with the pressure of steam upon the valve-stem area, quickly shut the port, and the steam is allowed to expand from that point on to the end of stroke; the higher the tappets are permitted to rise, by the elevation of the gauge-plate, the greater the ratio of expansion; the further they are depressed, the shorter the cut-off. As these engines are constructed, they are capable of cutting off steam anywhere between the beginning and three-quarters stroke; the latter limit is determined by the lead, and by the margin thought necessary to secure certainty of closure of the valve, when tripped, before the piston reaches the end of its stroke. To follow farther would not be likely to be of advantage, as the gain in the mean total pressure would be compensated by the loss due to a retarded exhaust. A safety stop-motion is combined with the governor connection, in such a manner, that if the belt breaks, or is thrown off its pulleys, the steam will be at once shut off, and the danger of accidents, such as sometimes occur with a run-away-engine, is avoided. The valves and seats on the exhaust side are both easily removable, from the outside, have outside connections, and are readily adjusted without interference with the steam side.

These engines have been, next to those of Corliss, the
pioneers in the movement, during the past generation,
toward economical working of steam. An engine built
upon this plan, substantially, by the firm of Thurston,
Gardner & Co., nearly a quarter of a century ago, from
designs prepared by Mr. E. D. Leavitt, Jr., for a well-
known Eastern mill, had steam-jacketted cylinders, 26½
inches in diameter, 5 feet stroke of piston, made 50 revolutions per minute, with steam at 100 pounds pressure in
the steam chest, and, on trial, worked down to a consumption of 1.98 pounds of coal per horse-power and per
hour; the guarantee was 2 pounds. Its fly-wheel, designed
by the writer, who was then just out of college, weighed
about 20 tons, was fitted up as a mortice gear, with cut
hickory teeth, and was given extremely small side clearance; the motion of the engine was so smooth, however,
that the presence of the gear was hardly noticeable. This
engine was fitted with the gridiron slides, as in the above
illustration; they were driven by sliding cams, thus ob-
taining a rapid opening and closing of the exhaust, and a
slow movement while in the intermediate position, with the
port either open or closed. This was a remarkably good
piece of work for that time, and has not often been
excelled since.

This engine, like other engines with drop cut-off valve-
motion, is not adapted to such high velocity of rotation as
to permit it to work safely at the speed of even the largest
and slowest of the modern "dynamos;" but, belted to the
machine, it will give as great economy, and as great perfection of regulation, as engines of the preceding class. It

is evidently so arranged that no load is thrown upon the governor, and the effort to detach the steam valve is, therefore, not liable to cause any oscillation in the cut-off gear, or variation in the speed of the engine. In all these engines, the difficulty met with by the designer is, not to secure this independence of the governor from the action of the valve-gear, but to prevent the irregularity which comes of the oscillations of the governor itself. The dash-pot attached to the governor, or, sometimes, a friction mechanism, prevents such irregularity.

This valve-gear does not as conveniently adapt itself to the vertical engine as some others, but one of the first engine-cylinders ever designed by the writer, was built with this gear, and was set vertically. It gave perfect satisfaction, if the fact that it was never reported to the shop for repairs, so far as the writer has yet heard, may be taken as evidence of its successful operation.[1]

This engine was introduced over a quarter of a century ago, in the face of a strong competition from the Corliss engine—a fact which is, perhaps, the best evidence that it had merit—and by the same methods which Mr. Corliss had proved so effective. Guarantees were given of performance, and forfeitures were provided for in the contract; or else the agreement was accepted to take as payment the saving actually effected in a fixed period of time—usually from two to five years, according to the character of the machine displaced. One of these engines, with which the writer was familiarly acquainted through his indicator, and

---

1. This engine is still in use, after 22 years service, and drives a set of dynamos at South Webster Mass.

which displaced the rival engine on such a guarantee, has now been in operation 23 years, and is reported to be to-day still in perfect order. The engine referred to above as having given so excellent a performance, was put in under an agreement by which the builders agreed to forfeit $1,000 per ¼ pound that the coal consumption should fall short of the guarantee. The manufacture was interrupted for some years by an injunction secured by Mr. Corliss, after a suit brought by him for infringement, but was recommenced after the expiration of the Corliss patent, and has proved a successful enterprise, notwithstanding the fact that its constructors have depended, apparently, upon the performance of the engine itself for advertisement—a conservative system of doing business which few manufacturers adopt, at present.

All three of the great inventors and introducers of the modern American type of steam engine—Sickles, who brought into use the drop cut-off ; Corliss, who gave the stationary engine its now standard form, as well as devised his peculiar valve gear; Greene, who applied the principles of this system of working steam to the plug-tree form of valve gear,—are now still living. Mr. Corliss has acquired wealth, as well as fame; his predecessor and his rival, however, have attained less fame—much less than they are entitled to, and still enjoy all the advantages which poets ascribe to the possession of small means. Neither of them expects to be able to build a monument, in the shape of a great technical school, such as it is becoming customary for wealthy engineers, like Stevens, and Rose, and Stone, to erect.

Kyes-Woodbury Co.

The Greene Engine.

There are other engines belonging to the class here considered—the engines having a detachable cut-off valve closed independently of the motion of the valve-gear,—of which the space proposed for these articles will not permit description. Among these are the Wright engine, constructed by one of the oldest and best known designers in the country; the Brown engine, a machine which has been extensively adopted for driving mills in New England, and is famous for the excellence of its workmanship and finish, as well as for its durability and efficiency; the Fitchburg engine, and others.

IV.

Engines Capable of Direct Connection.

THE PORTER-ALLEN ENGINE.

THE essentials, in the construction of the steam engine with a view to the economical production of power, as has been seen in the introductory part of this series of articles, include special provision against loss of heat and condensation of steam, at entrance into the steam cylinder, by the action of the metal surfaces to which it is exposed on all sides at the beginning of the stroke. One of the methods of securing this economy in the working of steam, has been stated to be the driving of the engine up to the highest safe velocity of piston, and giving it a maximum speed of rotation. The time allowed for condensation of each charge, and for the necessary change of temperature preceding such condensation, is thus reduced, and the amount of steam condensed being thus made a minimum, in any given time, the percentage of loss of the increased quantity of steam worked off by the engine becomes the least possible. The engine does a greater amount of work, and is subject to less loss. Thus the work to be done being fixed, it is done by a smaller, and, other things being equal, a less costly engine, and at the same by a more economical machine.

Although this seems a sufficiently simple and axiomatic philosophy, and although the general tendency of practice in steam engineering had been plainly in this direction for many years, these points had not, up to a comparatively recent time, been recognized by constructing engineers, and their progress had been slow and difficult. The older firms who were engaged in the building of what were then called "expansion engines," were the first to detect this movement and its cause, and they led off, in a very conservative way, toward the construction of faster engines. The firms already mentioned as leading in the movement toward correct practice, came up to speeds far ahead of those common among other makers, and secured an advantage that was sufficient to prove unmistakably that they were in the right track. They did not, however, modify their designs in any great degree, with a view to adapting them to very high speeds. Their valve-gears were not of a kind well fitted to high speed of rotation; the builders, were themselves disinclined to accept the risks undeniably attendant upon rapid change in this direction, and the public to whom they looked for a market were not educated up to such a point as would make it safe to attempt to go on very rapidly. A rather slow engine, with its comparative immunity from risk of serious accident in case any little derangement should occur, and with its greater durability under the ordinary conditions of use, was, by the great majority of designers, builders, and steam users, thought a far better investment than a fast engine, however well adapted to the radical illustration of a very interesting, but apparently impracticable, philosophy.

The first man to take up this matter with a will, and with a faith and a determination that were equal to the task, was Mr. Charles T. Porter, a young lawyer turned engineer, and Mr. John F. Allen, when the writer first knew him, a skillful mechanic, who was showing the natural bent of a real inventor, in the production of new devices, while engaged in the management of some of the best engines of 20 years ago. The valve-gear of the Porter-Allen engine, is the invention of Mr. Allen, and its governor and general · arrangement are due to Mr. Porter. It was Mr. Porter, also, who, by his courage, persistence, skill in business, and general good sense and management, finally, after years of struggle to secure good construction and workmanship, brought the engine into use in spite of every discouragement, whether due to circumstances, to direct opposition of competitors, or to public sentiment in favor of conservatism.

There are some interesting problems which present themselves to the engineer who attempts to design an engine to be operated at very high speed—problems which are by no means easy of solution, except to the boldest of innovators. One of these points of difficulty has already been considered. When the speed of revolution is increased, it is evident that a limit must sooner or later be attained at which the drop cut-off must be exchanged for some "positive motion" gear. But the various forms of such gearing familiar to engineers when Messrs. Porter and Allen became acquainted with each other, years ago, the still common three-ported valve, such as is used on locomotives, the Meyer valve with its cut-off valve on the back of the main valve, and kindred devices, were not adapted to the conditions sought by the engineer looking for a good system of expansion. They were simple and inexpensive, and could be used at any practicable speed of engine; but they did not always give a satisfactory distribution of steam. They usually produced a retarded steam supply, a "throttling" of the steam at the point of cut-off, which was not at all such as would satisfy the engineer familiar with the prompt action, and the "sharp corners" of the indicator diagram from the class of engine then taking the market. The dependence of the several parts of the motion upon each other was another objection to these devices, and the load which they threw upon the governor was a fatal defect, as the governor was then arranged and connected. Mr. Allen's invention placed in the hands of Mr. Porter just the device that he needed to carry out his idea of a fast engine.

This arrangement consists of a single eccentric driving a link motion to operate the steam valve and to work the exhaust at the same time. The link is controlled by a Porter governor, and is so connected and driven that the gear may be readily and quickly adjusted by the governor to any desired point of cut-off.

The eccentric and link are shown in the next illustration. The eccentric is set on the shaft in such a position, that its motion is co-incident with that of the crank. The link is a slotted curved arm, forming one piece with the eccentric strap, pivoted at the middle on trunnions sustained by an arm rocking about a pin set in the bed of the engine. The upper end of the link carries a pin, from which a rod leads off to the exhaust, which is driven without variable connections. The link-block is fitted to work in the slot of the link, from the end nearest the exhaust rod pin, down to the point opposite the pivotal point at which the trunnions are set. When it is at the upper end, the throw of the valve is a maximum; when at the lower point, it is a minimum. As the link-block is moved up and down in the slot, the motion of the valve is varied, and the ratio of expansion correspondingly altered. By an ingenious adjustment of a still more ingenious form of valve-motion, it is thus possible

THE ALLEN LINK.

to obtain a valve movement of perfect precision at all speeds, and on both the forward and the backward stroke, with a quicker closing action, as the cut-off is later. The steam is. allowed to enter the cylinder, at nearly boiler pressure, almost up to the point of cut-off, and the expansion line is a smooth curve very nearly from the junction with the steam line.

This form of indicator diagram has been usually considered peculiar to the class of engine described in the preceding articles. In this case, the diagram is nearly as sharp in the corners as those from a drop cut-off engine. The range of expansion is from the beginning of the stroke to about five-eighths.

There are four valves, as shown in the next Fig., which is a section through the steam cylinder showing valve, ports, and general construction. The two valves at the upper side of the cylinder are the steam valves; the lower are the exhaust valves. This section is, however, horizontal, the valves being set on their edges at either side of the cylinder. The exhaust valves are so placed as to drain the cylinder of any water that may have entered with the steam, or may have been produced by internal condensation. Both sets of valves are so made, and set, as to be well balanced, and so as to be capable of having the wear taken up when it occurs. The steam valves are provided with packing plates, which are adjustable by hand, to make them steam tight, as well as to secure a perfect balance. Each valve is placed in a separate valve-chest, and can be independently adjusted. Each valve opens four ports; each is so set, that it is actuated by a rod in the line of its own centre; and all are thus rendered but little liable to either wear or leakage.

The rock-shaft arm on the intermediate rock-shaft, seen in the large Fig. between the eccentric and the steam valve stem, assists in securing the quick opening and closing motion essential to a satisfactory distribution of the steam.

The features which have now been described, are not

necessarily distinctive of a "high speed engine." A positive motion valve-gear, and a good steam distribution, are
desirable in such engines, and the first point is, in fast running machines, an essential requisite; but the Allen engine,
so far as it has been described, may be as well considered
a slow as a fast engine. There are some details, to which
've are now to turn our attention, which are essentially and
peculiarly characteristic of the class to which this machine
is assigned. Among these points are the strength and rigidity of parts which distinguish such engines; the great
nicety of fitting; the excellence of all material in every
part exposed to the straining action of inertia, and the
minor but yet important modifications of details to adapt
them to service in a machine, in which the slightest play in
joints or bearings will be certain to make trouble. The
bed is of peculiar design and is enormously stiff and solid,
especially in those parts which take the stresses of the reciprocating pieces. It is broad and deep, with the line of
thrust of piston rod carried close to its surface between the
guides, and with a box form which gives great resistance to
forces tending to twist it.

The steam cylinder is secured to the bed by the end, a
construction adopted by Corliss many years ago, and one
which gives all desirable strength, with freedom from those
strains which come of connection of two large masses at
different and constantly varying temperatures. The whole
of its exposed surface is covered with lagging to prevent
loss of heat by radiation. The main journal boxes are
made in four pieces, and are set up by adjustable wedges,
so set as to avoid the springing of the shaft that is some-

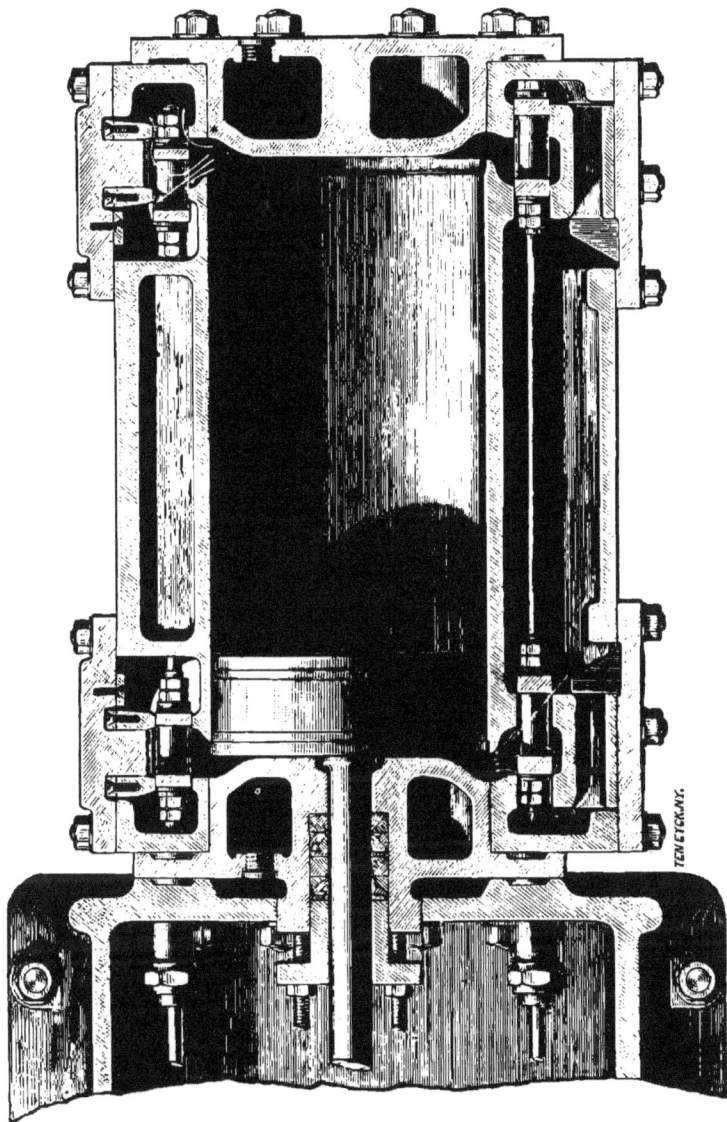

STEAM CYLINDER (SECTION).

TEN EYCK N.Y.

times found to occur with a less effective arrangement. The main-shaft journals, and the journals of the crank-pins, are made with especial care, skillfully ground to size and form, and nicely finished before the engine is assembled. The pin is always of "mild" steel, carefully case-hardened to give it a surface that will wear well and will not "cut."

The provisions for lubrication in such engines are not the least important of its details. The engine presents some neat devices in this respect which we have not space to describe.

One of the most remarkable and interesting of the features, which especially adapt this engine to great speed of rotation, and one, the developement of which, in its theory, as well as in practice, is due to Mr. Porter, is a peculiar adjustment of weight of moving parts to the equalization of stresses on the line of journals between the piston and the crank-shaft. When the steam is allowed to follow the piston only to some point early in the stroke, the ratio of expansion being made, as is usual, between three and five, the rapid fall of pressure, during expansion and up to the end of the stroke, causes a very great variation in the effort exerted upon the crank-pin and other journals. As the maximum pressure occurs when the crank is passing the centres, and while the work done usefully is, in consequence of the slight travel of the piston, very little, and as, at the same time, the considerable movement of the pin under this pressure causes a considerable loss of work by friction, and as it is advisable to secure a uniform effort producing rotation, it is evident that it is desirable to find a method, if possible, of equalizing the pressure throughout the stroke without sacrificing the advantages of expanding the steam. The action of inertia in the moving parts is made by Mr. Porter the means of securing this result.

At the beginning of the stroke, the inertia of the piston, its rod, the crosshead, and to a certain extent the connecting rod, all reciprocating parts, causes them to offer a certain resistance to the accelerated motion which they are compelled to take up. This resistance becomes less and less up to zero at half stroke, the point at which their velocity is a maximum. Passing this point, they are rapidly retarded, and this same property of inertia causes them to offer a resistance to retardation, which resistance now is felt as an impelling force at the crank-pin. Thus, the effect of the presence of these heavy masses in the line of connection, produces a reduction of pressure upon the pin at the commencement, and an increase of pressure at the end of stroke. But, in consequence of the varying action of the steam producing an excess of pressure at the beginning, and a deficiency of pressure at the end of stroke, we may combine these two effects, and the result is a comparatively uniform load upon the crank-pin throughout the stroke.

This compensation is capable of being, in many cases, very nicely adjusted by properly proportioning the weight of the reciprocating parts. As engines are usually proportioned with a view to strength of parts simply, the piston, crossheads, and rods are too light to be of much service in this way. Mr. Porter adopted the plan of making his piston and crosshead of such weight that the equalization of pressures should be the most complete possible, and this involved making them decidedly heavier than they are made in common practice, even when his engines were driven up to a speed which had never been before attempted in stationary engine practice. It is evident, however, that at some higher speed, the weights of these parts, as proportioned for strength simply, would be sufficient to give this desirable adjustment of the load on the crank-pin. There is no reason to suppose that this, which would seem to be a natural speed of the steam engine, may not be at some future time attained.

An interesting fact in this connection, is that Mr. Porter, although not professionally a mathematician, or educated as an engineer, first worked out the relations of these forces by a simple process, and applied his results to his practice, and that, subsequently, at his request, a distinguished mathematician, Dr. Barnard, President of Columbia College, attacked the problem by the methods of the higher analysis, and revealed the laws involved, and verified completely the work of the engineer.

The large engraving on page 49, represents one of three pairs of engines in use at the Willimantic Linen Company's mill. They are 11½ inches in diameter of cylinder, and 16 inches stroke of piston; they make 350 revolutions per minute. This is not considered a high speed for these engines, however. A considerable number of these engines have been used in the electric lighting service of large cities, to which service they were the first to be adopted for driving large dynamos directly connected. Under some conditions, it is found that the weighted governor is too sensitive, or too much affected by inertia, to give perfect regulation. For such cases, Mr. Porter has designed an isochronal governor, which is free from this cause of variation.

Engines of this class have many advantages, consequent upon their high speed; they are, other things being equal, more economical in the use of steam; they can be given a very much smaller fly-wheel; they have, in consequence of the enormously reduced weight of wheel, less friction; they are more easily held to their speed by the governor; they are less subject to variation of speed between beginning and end of any one stroke; and they are usually less troublesome and expensive to connect to the load than slow running engines. These advantages are common to all classes of engines, as they are driven up to high speeds; the class here considered is simply better fitted to realize these advantages than the older forms of engines, because they are especially designed for high speed. The objection to the "high speed engine," is the increased risk of wear, and of accident due to their rapid motion, and especially the risk, that when accidents do occur, as they will now and then in the best regulated establishments, they may be vastly more serious than with engines working at ordinary speeds. The object of the precautions which are taken by builders of fast engines, are all directed to meeting this contingency, and to making their machines safe against accident. These precautions are seen to be the strengthening, and especially the stiffening, of all the parts exposed to the stresses due to the action of inertia in the reciprocating pieces; the adjustment of all parts to each other in such a manner as to avoid spring; the use of the best material; an effective system of lubrication; and the securing of the most perfect workmanship.

Watt once congratulated himself that he was able to get

PAIR OF PORTER-ALLEN ENGINES.

a steam cylinder that only lacked three-eighths of an inch
of being truly cylindrical; the builder of the "high speed
engine" of to-day works to the thousandth of an inch, in
longitudinal measurements, and gets his cylindrical journals
exact to the twenty thousandth, perhaps to the fifty thousandth of an inch, a quantity which can be detected by a
good workman. The contrast illustrates well the progress
of a century in accuracy of workmanship where nicety is
required. Such nicety, only, can make a fast running engine safe; such accuracy *does* make it safe, and such
engines now do their work uninterruptedly, year in and year
out, and are found to require no more than that ordinary
care which all engines are expected to receive.

A Porter-Allen engine, from the "Southwark Foundry,"
supplied power to the Weston, Edison, and the Thomson-
Houston Electric Light Companies at the Railway Exhibition at Chicago, May and June, 1883.

THE "BUCKEYE" AND "HARTFORD" ENGINES.

THE engine last described was a long time alone in the
field as a "high-speed engine." The principle represented by its designers was recognized as correct by every
intelligent engineer, and it was admitted that the fast engine,
other things being equal, would prove the most economical
in its expenditure of heat, as well as in its efficiency as a
machine subject to friction. But builders were not able to
bring themselves to accept what seemed to them the risks
incident to high speeds. The pioneer in this new field was
not altogether successful for a time, and it seemed to be

certain at one time, that the engine, despite the pluck, the persistence, and the skill of its indefatigable promoter, must retire from the market.   But no discouragement could quite destroy confidence in this engine, which had become the embodiment of the most recent phase of progress.   Gradually, one difficulty after another was overcome; parts were strengthened and given satisfactory proportions; the materials were improved and the workmanship of the machine was made as nearly perfect as the best tools, handled by the best workmen, could make it.   A little gain was seen each year, and, after a time, it was seen that the new class of steam engine had "come to stay."

One of the first engines to come into the field after this period of doubt had closed was built by an enterprising firm of Western manufacturers.   This was the " BUCKEYE ENGINE," designed by Mr. J. W. Thompson, and built by the Buckeye Engine Co., at Salem, Ohio.   The engine did not start as a radical competitor of the pioneer engine; but it was from the beginning, a moderately high-speed engine.   It was fitted with a positive motion "automatic" valve-gear and a balanced valve, and had a stability and an excellence of workmanship that made it safe at fast speeds; while the peculiarities of its construction were such as gave it a very high place as an economical machine.   It was capable of meeting in competition the best engines of the day.

The form given the larger sizes of this engine is seen in the preceding Fig.   The general arrangement is not essentially different from that of the Corliss engine, which has been described in earlier articles.

The cylinder is carried on a pedestal, as is the latter;

the frame consists of a girder uniting the cylinder and the main pillow block and carrying the guides; the crank-shaft end is carried by another pillow block. The main frame is, however, supported by a strut which is now usually seen in other engines, and which takes the load tending to spring the girder under the guides. The construction of the cylinder, and the arrangement of the valves, is shown in the next Fig.

The live steam is taken into the steam-chest at A, passes through the passage, a, a, through the openings, D, D, into the box-shaped valve, B, B, and thence through the ports, b, b, into the cylinder, as the ports in the cylinder are alternately brought opposite those in the valve. The cut-off valve is formed of two sliding plates, C, c, connected by rods, C', and sliding on seats formed on the inner, or working, side of the main valve, so as to cover the main steam ports alternately, and at times which are determinable by the governor. The stem, g, driving this valve, passes through the main valve stem, which is made hollow for that purpose. The cut shows the steam entering the cylinder at the left, and the cut-off valve just beginning to slide over the port, while the exhaust is taking place at the right, *past the end of the main valve*, through the chest, and around to the exhaust pipe seen partly dotted at F. At e, e, are seen two "relief chambers," which receive live steam from the steam valve through holes, f, f, and thus balance the valve at a time when the pressure on the seat caused by the then excessive area of the balance openings, D, d (which openings must be made sufficient in area to produce a slight pressure of the valve on its seat when the tendency to lift

the valve from its seat is greatest), is overbalanced.   These
holes only fill when this relief is needed.   The equilibrium
rings, *D*, *d*, seal the joint between the valve and the dia-
phragm separating the steam-chest, *a*, *a*, from the exhaust-
chest *F*.

The governor is of a type that has not been seen in the
engines previously described.   It is shown in the following
illustration, page 68.

In the common "fly-ball governor," the two balls revolve
about a vertical spindle, to which they are attached by a
pair of arms in such a manner that they may take any position that the resultant action of gravity, centrifugal force,
and the pull on the supporting arms may give them.   A
defect common to all governors of this class is that the
force tending to pull the balls downward is perfectly uniform.   Gravity never changes at any one place.   The position taken by the balls, at any fixed speed of engine, is
always the same; the connection of the balls with the regulating mechanism, is one which always preserves a fixed relation between the position of the governor balls and the position of the regulating apparatus.   Thus it happens that the
engine can never be kept precisely at speed, unless the speed
is such as will give the governor exactly its normal position
and, at the same time, such that the valves shall supply just
the normal quantity of steam to the engine.   With reduced
steam pressure, the engine drops to a slightly lower speed,
and runs at that speed instead of the proper number of
revolutions; when the load decreases, the engine runs at a
little higher speed than is intended; and no method of
attaching that form of governor can give absolutely uni-

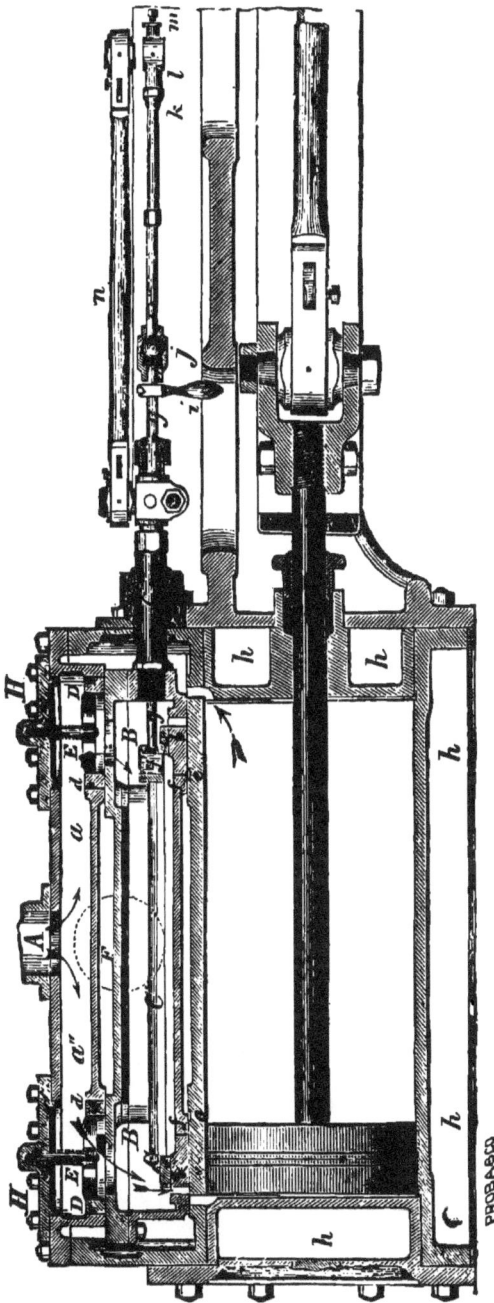

SECTION OF CYLINDER AND VALVES.

PROBA&CO

form speed. If, however, we can substitute for the action of gravity, a force which can be made to vary with change in the position of the balls, in such a way that the variation in the opening of the throttle, or in position of the point of cut-off, shall go on until the engine comes to speed, irrespective of all other conditions, we shall have what is known as an "isochronous" governor, and shall be able to get the correct speed, whatever changes occur in steam pressure or in load, provided that there is steam enough to drive the load at speed with the least expansion for which the engine is designed. Such an adjustment can be made by substituting the tension of a spring, properly set, for the action of gravity. The form of governor here illustrated is, or can be made to be, of this class. It simply requires that the spring tension shall be given a certain easily determined relation to the effort of centrifugal force.

A governor of this character, when well made and adjusted, will open the throttle valve, or will increase the ratio of expansion, as the steam pressure diminishes or as the load is increased, and will continue to move in the proper direction, indefinitely, or until the machine comes to speed, or until the engine is doing all that it can do. In the governor here used, two levers are set on either side the crankshaft, in a frame or a pulley to which they are pivoted at *b, b.* These rods carry weights, *A, A,* which may be adjusted to any desired position by means of the bolts seen in the cut. The outer end of each rod is linked to the loose eccentric, *C, C,* by the rods, *B, B,* and is controlled by the springs, *F, F,* which resist the effort of centrifugal force tending to throw the weights outward. As the weights

swing outward or inward, as the one or the other of the two
opposing forces predominates, the eccentric is turned on
the shaft in such a manner as to give the valves that motion
which is necessary to produce the proper distribution of steam

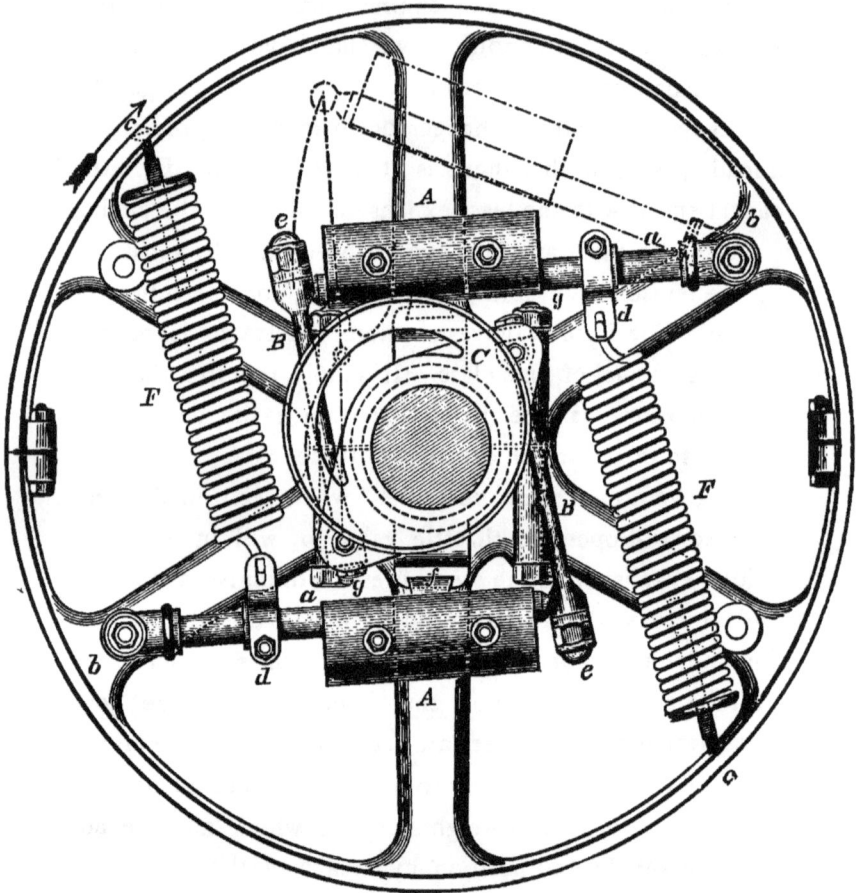

GOVERNOR.

to bring the engine to its speed. The adjustment of this
regulator to its work is easily obtained by the shifting of
the weights along the levers, or by increasing or diminishing
their amount, as is found necessary.

BUCKEYE AUTOMATIC ENGINE.

This governor is adjusted for an engine moving in the direction of the arrow. To adapt it to an opposite motion, the pins, *b, b,* are shifted to the other set of arms which are shown having bosses for their reception. Wooden buffers check the governor at the extremity of its range of motion.

The range of expansion, as determined by the governor in this engine, is from the beginning up to two-thirds stroke.

The engine has many interesting peculiarities of construction, in its details, which space will not permit us to consider.

The HARTFORD ENGINEERING COMPANY, who are building this style of engine, make a form of bed which is somewhat similar to that designed by the makers of the Porter-Allen engine, but which is particularly solid and graceful in appearance. It is seen on the opposite page.* This firm, as well as the original makers of engines built under Thompson's patents, endeavor to secure in their engines, great weight in the parts in which solidity is important, such large area of bearing surfaces as is essential in these engines, moderately high-speed of piston and of rotation, a steam pressure, usually of about 80 pounds per square inch, and adopt a ratio of expansion for their non-condensing engines, of from four to five. Their table of powers of their standard sizes is based upon estimates for steam at 80 pounds and a cut-off at one-fourth. In construction, these engines are carefully made with all joints

---

* The first designer to carry the line of the steam cylinder along the surface of a "box-bed," and thus to secure maximum vertical and horizontal stiffness in this manner, so far as the knowledge of the writer extends, was Dr. E. D. Leavitt, Jr., who made such an arrangement in engines, in the design of which the writer assisted, as early as 1860.

scraped, and all pins, and all journals also, ground with scrupulous care.

The method of regulation is, as has been seen, quite different from that practiced by the older standard makers. It is subject to the objection, that as the regulator has thrown upon it the duty of altering the position of the eccentric, the load so brought upon it may make it less sensitive and less effective in regulating the speed. This conclusion, which is that usually held by the older engineers in the profession, seems to be contrary to the fact; although, when comparing the older kinds of engines, it is fully sustained by the superior regulation of the engines of the " automatic " class. The fact, now familiar to every engineer accustomed to the management of electric lighting machinery, that engines having regulators of the class to which that under consideration belongs are capable of giving a good regulation, even when directly connected to the dynamo, is sufficient proof that such a system of regulation may be able to do perfectly satisfactory work. The frictional resistance of the system, while in motion, is not a matter of importance; as in any system in movement, and subject to jar, the friction is practically eliminated and every part assumes the position that it would take in a similar system free from friction. The action of the resistance of the valve, so far as it is transmitted to the regulator, probably acts to hold the regulator fast during the period of its action, leaving it free to move into any new position, corresponding to the speed of the engine at the instant, without hindrance during the remainder of the time.

All of these fast-running engines will be seen to have

Hartford Standard Engine.

shorter strokes of piston than is customary with the earlier types. One reason which has guided their designers to this proportion is that the loss by internal condensation becomes less as the steam is given less time to discharge its heat, and hence high-speed of rotation and short strokes are adopted. The best proportion of stroke to diameter of piston, the number of revolutions in the unit of time being fixed, is easily ascertained by a very simple investigation. It is found to be two to one. This is about the proportion generally adopted in these engines. Many engines are, however, given a ratio of 1 1-2 to 1. The shorter stroke has the great additional advantage, the speed of piston being the same, of giving a less costly engine to build. The proportion is sometimes dictated partly by the character of the work to be done; thus, in driving the dynamo directly, the velocity of rotation must be very great and a short stroke becomes advisable—the shorter as the speed is higher. In such cases, therefore, engines are often made with even shorter strokes than considerations of "efficiency" alone, would dictate.

Reviewing the construction of this engine, it is seen that it is distinguished from those which have been already described, by its peculiar balanced valve which can be proportioned to take any desired part of the steam pressure, leaving, if properly adjusted, just enough on the valve to hold it with certainty to its seat and to secure a little wear to give bearing and fit between valve and seat, that this valve is arranged to take steam through, and to deliver steam outside, the shell; that it has a system of perfectly flat wearing surfaces, and a positive movement of in-

variable extent, and thus is not liable to the formation
of shoulders on seat or valve; that its clearance is so
small that it is easy to counteract any ill effect, ordi-
narily due to that cause, by moderate compression; that it
has two ports and thus possesses such advantages as may
be claimed for that arrangement; that the governor is driven
by a positive connection with the shaft on which it is set;
that, as the cut-off is adjusted by the motion of an eccentric, the ratio of expansion is the same at both ends of the
cylinder and that it possesses the advantage, common to all
engines having a positive motion valve-gear, of being unre-
stricted in speed.

Many of these engines are already in use driving electric
lighting machinery.

THE CUMMER ENGINE.

ALL of the class of engines now under consideration
have been seen to differ radically from the engines
previously described (as not well fitted for direct connection to the dynamo), and to have a number of characteristic points in common which especially fit them for use in
direct connection. This latter class of engines, however,
exhibit some differences among themselves which are important and very interesting to the engineer and the user
of steam power.

The engine last described will have been seen to differ,
in a very notable way, from that which immediately pre-
ceded it. The latter had a system of valves that differed
from the former no less radically than did its system of
regulation. We have now to study an engine which re-

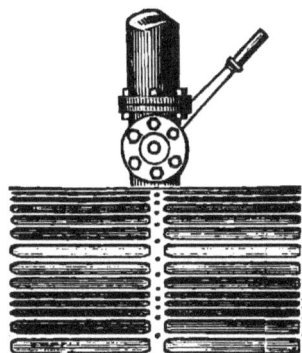

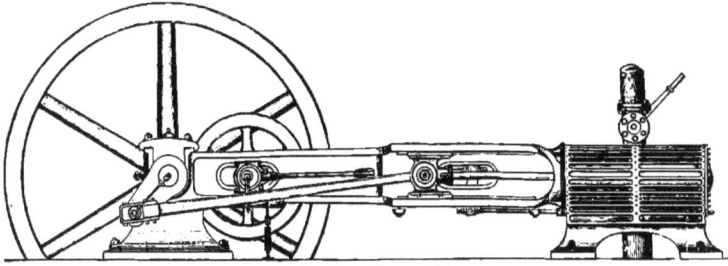

THE CUMMER ENGINE. "C"

sembles the last in its general features—the use of a cut-off valve riding on a seat formed upon or in the single main valve, a system original in principle with Meyer, an engineer well-known, years ago, in Europe, and the use of the peculiar form of governor which adapts itself to a position on a horizontal or on an upright shaft with equal facility. This engine, however, has some curiously interesting and ingeniously contrived points of construction which, as well as its performance, make it well worthy of attention. This, the "Cummer Engine," is illustrated in the engravings to be described below.

The Cummer Engine Company makes a number of different forms of engine, using various kinds of valve-gear and different forms of regulator and of engine frame; but the style with which we are here principally concerned is that which is best adapted to driving a load at high speed with great economy and with the most perfect regularity.

The general form of the engine, as shown in the Fig., on page 74, is very similar to that of engines already described. It has the "girder" frame, or bed, is well supported at each end, has a firm and substantial connection in the line of thrust and pull between cylinder and crank-shaft, and provisions for lubrication especially fitted to give safety at high rates of speed. A modified form of bed is seen in the next illustration, in which one of the engines designed for the highest safe speeds is shown. In this engine, the frame is made with a pedestal cast upon it directly under the guides and extending under the whole length traveled by the crosshead, thus giving absolute stability at the point at which cross strains are most severe and most productive of injury.

The cylinder overhangs, unsupported, at the back end of the frame. No support is there needed, however, as no appreciable vertical stress occurs there. This engine has the same valve and gear, and the same form of governor as is used in the preceding style of machine. In this latter form of engine, the crank is replaced by a disc, an arrangement which enables the builder to effect a more perfect balancing of the reciprocating parts than can well be obtained with the ordinary form of crank. The rigidity of this form of engine is seen to be as essential a feature as in those which have been previously described. The box girder gives this stiffness in a very satisfactory manner.

THE CUMMER ENGINE. "B."

The main guides are flat, and are fitted with removable faces which can be readily repaired or replaced, when worn or "cut," at small cost of time and money. The crosshead is a compact, strong casting, having bearing surfaces extending well out under the pin, and under the piston-rod socket, as well, and it is therefore not likely to cause those awkward accidents, due to springing the piston rod at this connec-

tion, which have proved so costly in less well designed engines. The gibs which take the wear are removable and
adjustable. The main bearing is fitted with four-part boxes
of babbitted cast iron, the side pieces so arranged that they
may be set out to a bearing as they wear. All the details
are in accordance with standard practice in this class of
engines, and description is not called for here. It may be
safely assumed that this is the case in any successful engine,
as good workmanship, the best materials, and a strong system of connections, are essential pre-requisites to even the
beginning of success.

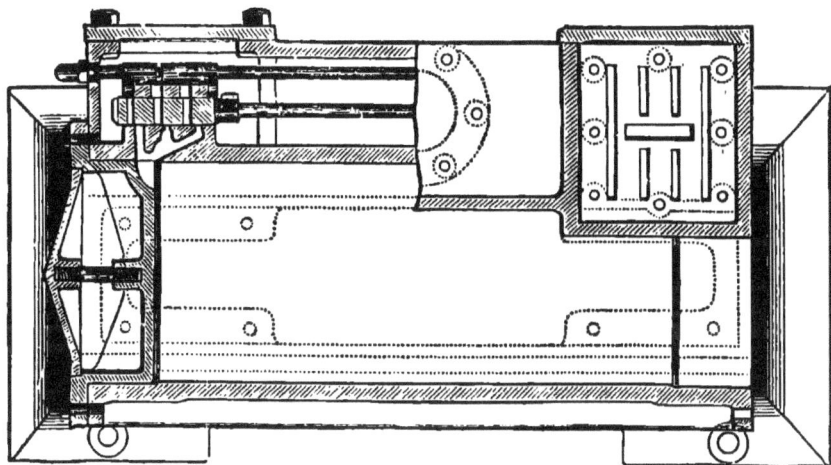

CYLINDER; STEAM VALVES.

The valves and the valve-gear of the Cummer engine,
as has been stated, belong to the " Meyer system " and consist of a main valve with the cut-off valve riding on the
back of the main. There is this difference, however, between the gear of this engine and others of the same general system : that here we find a separate system of exhaust

valves which are worked independently of the steam valves, and thus leave the induction and eduction motions entirely free to be adjusted as the designer, the constructor, and the user, may desire. The preceding engraving shows the disposition of the valves in the cylinder casting, and the larger cuts exhibit the method of driving them. The section of the cylinder, above, is made horizontally through the steam valve chest, and shows the main valve in section, with the cut-off valve riding upon it. At the left is a section so made as to exhibit the exhaust valve seat. This is made removable. It will be noticed that the valves are of what the engineer calls the "gridiron" pattern. They are so made, with their several ports, to obtain a free opening with small movement and reduced friction of the valve. The writer has found this device a decidedly advantageous one, and it has been used by some of the most successful designing engineers of his acquaintance. The more numerous the ports, the less the travel required for the valve, the smaller the steam chest space demanded, and the less the load on valve-gear and governor, usually.

The next illustration represents the same parts of the engine as seen from the side, with valve-chest bonnet removed at one end, and a section made opposite the supply pipe to show the passages and valve-rods. These rods are driven by the main eccentric, the steam valves directly, and the exhaust through a rock-shaft. The cut-off valve is driven by a separate eccentric, as in the preceding form of engine, and this eccentric, like the preceding, is adjustable in position on the shaft by the governor. The engine is thus made "automatic" in its adjustment of the point of cut-off, and in regulation. Separate valves are seen at each end of the cylinder, and the "clearance" and "dead space" is thus reduced to a minimum. This last provision makes it possible to "cushion" the exhaust steam up to boiler pressure on the return stroke, and thus to secure a minimum waste by condensation on the opening of the steam valve for the succeeding stroke. Cushioning is not here limited by the steam side. The construction of the connecting rod, and the method of connection, are such that the wear of jour-

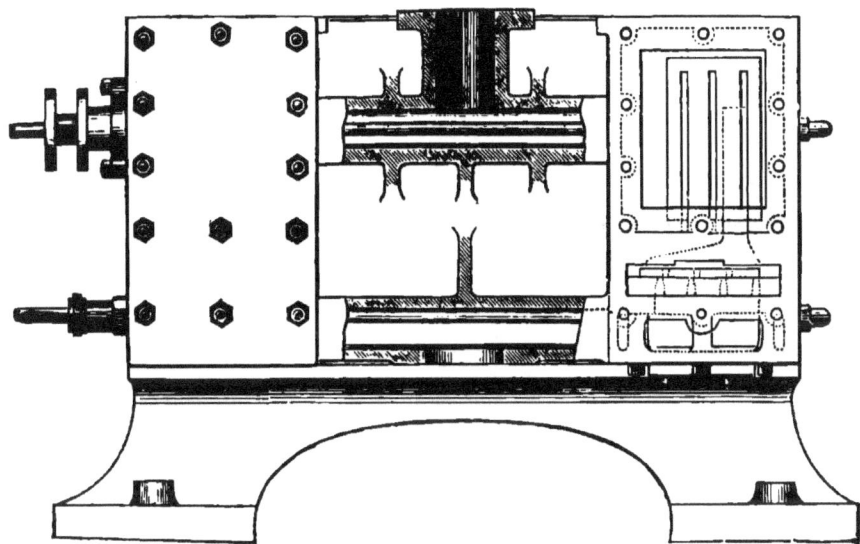

CYLINDER; ELEVATION AND SECTION.

nals and bearings may be taken up, in any case, without altering, to any observable extent, the position of the piston in the cylinder, and this permits small cylinder clearance, also. For the reason above given, the port spaces are made no larger than is necessary.

A comparison of this engine with others of its class will

exhibit one very peculiar feature, in which this engine stands entirely alone. The governor is carried on a "governor shaft" which is geared to the main shaft, and which has no other office than that of carrying the governor and the eccentrics. It is evident that so radical a departure from standard design must have been caused by the possibility, actual or presumed, of thus attaining some very important result. A little study shows plainly what this supposed advantage must be.

The necessity of providing for efficient performance at high speeds of rotation has been seen to have compelled the adoption of a positive motion valve-gear; the adoption of this gear led to the use of a powerful form of governor, directly attached to the cut-off eccentric; this, in turn, compels the use of revolving weights, turning in orbits lying in the vertical plane; this last feature, in turn, again made it necessary, apparently, to place the governor on the main shaft, and to meet the effort of centrifugal force by a counterbalancing action, which could then only be obtained by the use of steel springs set in the casings of the governor. But the use of springs is considered by many engineers to be so objectionable, that they would submit to some expense and inconvenience to avoid their application, if possible. The objections are that they are liable to changes of tension and of length while at work, that they never have a definite and calculable strength, that they are liable to break in most unaccountable ways, and at most unreasonable and unexpected times, and that the adjustment of a balance between the two equilibrating forces is often difficult and almost always unsatisfactory. These objections undoubtedly do to a certain

extent exist; but they as certainly are not as serious as is
often supposed. The writer has had a long experience

THE CUMMER GOVERNOR SECTION.

in this direction, both in the use and in the observation
of the steel spring for a wide variety of applications,
and has never yet seen reason to condemn them unre-

servedly. The principal objection which can be urged
against the governor of this class, as usually adopted for the
kind of engine now under consideration, is probably the
fact that it cannot be reached while the engine is in operation, and that change of speed is thus made impossible
except by stopping the machine and making changes in the
adjustment of the springs, then trying the speed again, and
again stopping to adjust, until the desired speed is exactly
attained, which disadvantage is shared by the older arrangement of governor.

The form of the Cummer governor, which has been designed to evade these objections to the use of springs, and
to secure certain special advantages, is shown in the above
illustration and in that which follows. As has been seen,
when studying the design of the engine as a whole, the
governor of the Cummer engine is of the same general type
as that of the engine last described; but it is mounted upon
a shaft, separate from, and driven by gearing from, the main
shaft. The governor shaft also carries the eccentrics, one
of which is loose on the shaft and is controlled, as to position,
by links from the weights of the governor as usual. The
governor is thus enabled to shift the eccentric forward or
backward and thus by changing its lead, to determine the
movement of the cut-off valve and the ratio of expansion.

There is nothing specially remarkable about this part of
the arrangement. The position of the weights is seen to be
determined, however, by a system of bell-crank levers which
connect the middle point of each weight with a vertical rod
and chain under the engine bed, and on this rod is carried
a set of weights which may be easily reached when the

engine is running. The bell-cranks within the governor casing, move a rod which passes along the centre line of the governor shaft and emerges at the left. This rod engages a large bell-crank at the end of the shaft, through which the load suspended under the engine is sustained. But one spring, and that a small one, is seen in the whole system. The centrifugal action of the governor weights, when at the inner limit of their range, is met by the weights on the scale pan, and the spring is only required to meet

THE CUMMER GOVERNOR.

the additional action of the governor weights when they fly outward, as the engine increases speed. The more nearly an equilibrium is maintained between the action of the flying weights and the balancing load, at the proper speed of engine and at all possible positions of the governor, the more perfectly "isochronous" does the governor become, and the more exactly will the engine hold its speed, under all variations of steam pressure and of load. With this governor, the weights on the pan can be increased or diminished at any moment, and to any desired amount, whether the engine is in motion or at rest; the isochronous adjustment can be effected as nearly as desired, and the speed of engine may, at any moment be altered, much or little as may be advisable.

This accessibility of the governor, and the disuse of heavy springs to control it, are the principal advantages of this form of governor. It has also some incidental advantages which are worthy of notice, although of less importance. The governor shaft is comparatively small; this permits the use of very small eccentrics; this reduces friction and load on the valve mechanism, and this, in turn, adds a little to the efficiency of the engine, as a compensation for the introduction of an additional shaft. The one spring used here is smaller than that needed for other governors of the same class, and is relieved from tension entirely at frequent intervals, and the periods of " rest " thus given it are likely to insure an increase in its longevity which may prove to be a point in its favor worth mentioning. It may sometimes, although certainly not frequently, occur that an engine may be required to work, at different times, at certain different, but fixed, speeds. In such a case, it is easy, with this engine, to find a set of weights which when in place, will give each one of these fixed speeds; the engine can then be, at any instant, brought exactly to either speed by hanging on the scale pan the right weight for the speed. The several weights can be kept at hand for use as required. Such an arrangement may be sometimes especially useful in electric lighting.

Several styles of the Cummer engine, other than those which have been described, are built for the market. Those which have been here illustrated are, however, especially fitted for such work as is the subject of this article. Both of the forms which have been described are well adapted to use in electric lighting plants, and are proportioned for high speeds; they are designed for nice regulation and are likely to prove durable, economical, and otherwise satisfactory motors. They are intended for steam pressures of 90 or 100 pounds per square inch, and their rated powers are based upon an assumed piston speed of about 400 times the cube root of stroke, as nearly as it can well be reckoned by the old method of James Watt—a speed more than three times as great as was thought best in the time of that great engineer. Even this speed is not to be considered remarkably great for engines designed and built, as are these, with especial regard to the requirements of high-speed motors. The steam pressures adopted are those generally regarded by engineers as, on the whole, the best for ordinary purposes, and are those beyond which the gain in economy by further increase becomes rapidly less with even the best engines. The point of cut-off is calculated, in estimates of power, to be at from one-fourth to one-fifth stroke, and, as a rule, nearer the first than the last figure. The best ratio of expansion for any given case is to be determined by a comparison of cost of fuel and steam supply with other operating expenses, at the place of operation.

The engine above described has been used, in many cases, to supply power for driving dynamos in electric lighting, and has an excellent record in that field, as well as in cotton

and flouring mills, which demand the most perfect possible
regulation.

One of these engines (16x36), at the Cincinnati Exhibition of 1883, was tested by the committee on electric
lighting apparatus and found to alter its speed but $2\frac{1}{2}$ per
cent., when the whole load, 124 horse-power, was thrown on
or off; it varied one revolution per minute with a change
of steam pressure of from 90 down to 50 pounds.

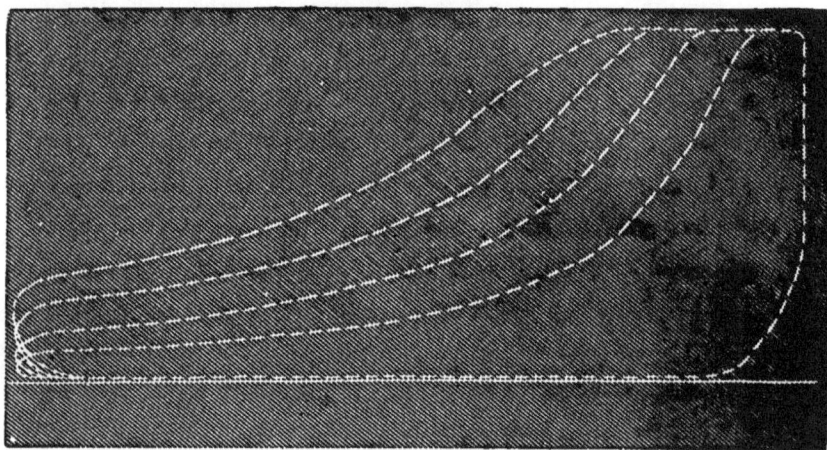

The indicator cards, of which copies are given above
as taken from this engine, show the method of distribution of
steam in engines with positive motion valve-gears, such as
are here considered as fitted for direct connection with
large dynamos, and for high speed generally.   The illustration exhibits a series of indicator diagrams taken from this
engine at points of cut-off varying from one-tenth to one-
third stroke.   It is seen that the steam lines are as straight
as those of a drop cut-off engine, very nearly up to the point
at which the effect of closing the cut-off valve begins to exhibit itself in the production of the expansion line.   The

expansion curve is very nearly that obtained by laying down the hyperbolic curve of Marriotte, and the exhaust is as clean and prompt as need be desired ; the back-pressure line closely follows the atmospheric line seen immediately beneath it, and the compression line at the right hand end of the card is quite as good as is often seen in the most perfectly proportioned engine with detachable valve. As the steam follows further and further, the sharpness of the corner between steam and expansion lines gradually becomes less, and the form of that part of the diagram approximates that found in the older forms of plain slide valve engine. For the most generally desired ratios of expansion, however, the form of the curve is satisfactory, and it is evident that the adoption of the positive motion type of valve-gear does not introduce any very serious loss of efficiency in this respect.

THE STRAIGHT LINE ENGINE.

REVIEWING what has been said in this section of engines capable of direct connection to the dynamo, it will be noted that the engines which have now been described have belonged to two classes, differing from each other in two very important respects. In the first, represented by the Porter-Allen engine, we find a form of engine especially, and very ingeniously, designed for high speed of rotation, fitted with four balanced valves, with the object of securing minimum " dead space," and maximum economy and ease of working, and controlled by a governor which differs from the older form introduced by Watt, by several useful modifications of design, and especially, by being loaded in such a manner that its speed, and, consequently, its power and sensitiveness in working, are greatly increased. In the second class, we find a valve-gear of the Meyer type driven directly by the eccentric, instead, as in the first class, through a link, and regulated by a governor riding on the main or the governor shaft, beside, and directly attached to, the eccentric. The features essential to a "high speed" engine are also embodied in the second, as well as in the first, class of engine.

We now come to the examination of a third-class of high speed engine, which differs as radically from the two preceeding as they from each other. In this new form of engine we find but a single valve which does duty both as a distributing and as a cut-off valve. A form of engine belonging to this class, with which the writer happens to be familiar, is that known in the market as "The Straight Line Engine."

This engine, so far as it is novel, is the invention of, and also is designed by, Professor John E. Sweet, formerly the superintendent of the workshops in which instruction in machine work was given in the Department of Mechanical Engineering of Cornell University—a position in which he became widely known as one of the most skilful and ingenious mechanical engineers in the United States— later a President of the American Society of Mechanical Engineers. The first of these engines was built at Ithaca for experimental purposes, by students under the instruction of the designer.

The Straight Line Engine has many interesting and novel

STRAIGHT LINE ENGINE.

points, which will bear much more extended study than they can be given in the small space which can here be allowed for the description of the engine. The problem, proposed to himself by the inventor, was to design an engine which, while consisting of the smallest possible number of parts, should, nevertheless, be economical in its use of steam, capable of the most perfect regulation attainable with any known device, strong and stiff in every part subjected to the working strains of an engine working at high speed, inexpensive in first cost, and durable as a simple engine can be.

This engine is shown in the accompanying illustration.

A vertical engine, which is shown at the end of the article, is also designed for all powers ; there seems no reason why it should not prove a good style for heavy work; better in some respects, in fact, than the horizontal engine.

The engine takes its trade designation from its peculiar form of frame, which is seen to consist of two perfectly straight diverging struts extending from the end of the cylinder directly to the two main bearings, thus carrying the line of resistance to the pull and push of the connections exactly along its own central line. No possible arrangement could give greater stiffness with the same weight of material. The whole structure is carried upon three points of support, as is the practice with "surface plates," which must, if possible, have an absolutely definite and invariable system of suports, to avoid the slightest danger of "spring." These points are under the main bearings, and beneath the steam cylinder. The two journals receive equal loads ; the crankpin is not subject to the deflecting forces met with where a crank is overhung ; danger of unequal wear of journals, and of springing the pin, is thus avoided very completely. The fly-wheel is placed in twin-form between the main bearings, and also serves as crank, thus making the best of cranks as well as balance wheel. This position of the balance wheel is one of peculiar advantage. By its action at this point, it intercepts heavy and objectionable stresses, which, in other engines, are transmitted to the main shaft ; and the reciprocal action of counterweights and equilibrating parts is thus only felt within a mass of metal, which can resist them with perfect safety, and without their being felt in the more sensitive parts of the machine. This arrangement renders the main journal less subject to springing under the loads transmitted through it. To secure better distribution of wear, the crank shaft is allowed some end-play. This end-play, together with the carefully arranged system of lubrication, are the best possible insurance against excessive friction and wear.

The steam cylinder has the appearance of the cylinder familiar to every one, as seen on ordinary plain slide-valve engines. Its valve chest is placed at the end nearest the crank, and the ports and passages are carried as in those engines. The valve stems have no stuffing boxes, but pass into the chest through unusually long and carefully fitted holes, in a hub, made about five one-thousandths of an inch larger than the rod inside the Babbitt metal bushing, for a length of six diameters or more. The hub is loose in the hole in the end of the valve chest, and is packed at the ends by a washer fitted on a flat seat on the inside. The piston-rod is similarly fitted.

The crosshead is a very long casting which overruns the

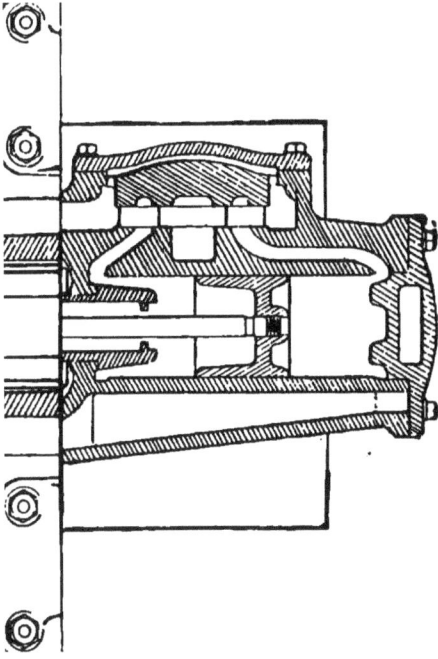

STRAIGHT LINE ENGINE.

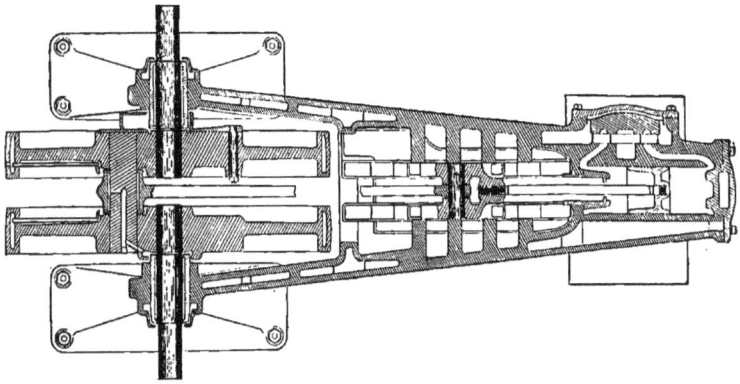

STRAIGHT LINE ENGINE.

guide at each end at every stroke, and thus is rendered safe against wearing to a shoulder. A pin subject to recipro-cating efforts in any part of an engine, whether it rotates, or carries a rotating or a vibrating piece, is apt. in time, to show wear on the two sides in line with the principal pull or thrust, and to lose its cylindrical form. In this engine, such wear is avoided at the crosshead pin, by cutting away the surfaces, which do little or no work, and thus securing overrunning surfaces, which are not subject to this distorted wear to so great an extent. Many other minor points invite attention, but they cannot be here considered.

The principal feature of this design, in connection with that phase of its work which is of especial interest here, is its valve-motion. The valve is a rectangular block, sliding between the seat and a coverplate; is shown in the engraving. Ports are cut through the coverplate, through the valve, and through the seat into the steam and exhaust passages in the cylinder casting, in the proper positions. These ports are double at the ends of the valve, and a single port of ample area is made through the middle of the valve.

The valve is what may be called a "piston valve" of rectangular section, the space in which it slides having, therefore, also a rectangular section, and permitting the use of a detached coverplate, which,while sustaining the pressure of steam that would otherwise come upon the valve, and thus making it a balanced valve, nevertheless allows any unusual pressure, occurring when the piston comes back to the compression period of its cycle, to raise it, and thus to permit the water which may have caused the pressure to flow back, and thus relieve the cylinder, and obviate all danger of forcing out the heads. The principal feature of this device is not new ; the writer handled such a balanced valve on marine engines, rated at above 5,000 horse-power, nearly twenty years ago, and found them, so far as his own experience went, perfectly satisfactory. This new application of the principle, however, embodies some new and interesting points. The valve cover is sustained on loose packing strips, which are free to close up upon the edges of the valve, and to take up wear as it occurs. The form of the plate, its domed top, is such as to give it great stiffness against the superincumbent pressure, and thus to prevent pressure on the valve itself in consequence of spring in the plate, and the ports are so placed as to prevent the cutting away of the faces and seats by the rushing currents of steam.

The valve and cylinder ports are not dressed out ; the casting is made so accurately that these edges can be left as they come out of the sand without loss of efficiency in the working of the valve.

The valve is driven by an eccentric, the motion of which is controlled by the governor, and the connection of which with the valve is effected by the peculiar system of linking, seen in the preceding illustration. The eccentric is so suspended from the disc, to which it is attached, that it may be thrown across the shaft by the action of the governor, in such a manner as to give the effect of the once common and well known "Dodd motion." It is carried on a lever, which is pivoted at one side of the shaft, while the governor rod is attached at the opposite side. The singular positions of the eccentric rod and the rockshaft arm enable the alteration of the throw of the eccentric produced by

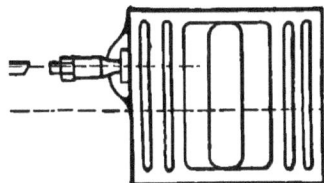

STRAIGHT LINE ENGINE —GOVERNOR AND VALVE GEAR.   ELEVATION.

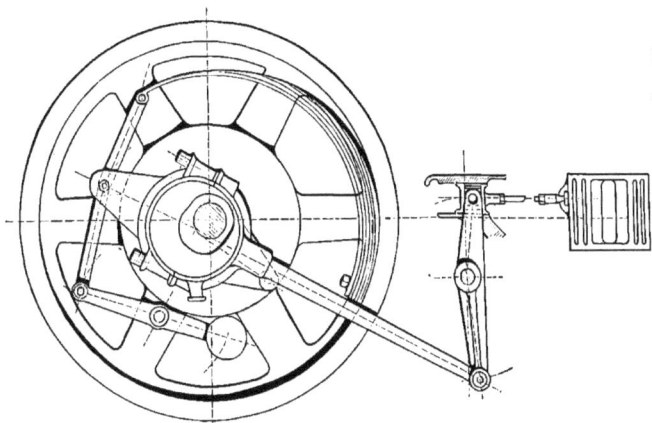

STRAIGHT LINE ENGINE.—GOVERNOR AND VALVE GEAR. ELEVATION.

the governor, to be effected without alteration of the lead of the valve, so that the steam may be admitted, at all times, at the same point in the revolution of the engine. This it does, since the line of the eccentric rod is, at the commencement of stroke, in line with the lever on which the eccentric is carried.

The governor is similar, in principle, to those which have been described as used on the last type of engine. It consists of a single weight, or ball, carried on the end of a lever which is pivoted, near its middle point, on one of the arms of the governor pulley, and connected to the spring, by which it is held under control, by a link extending across to the other side of the shaft to the end of the spring, which is there secured to the rim of the pulley. The action of the governor is substantially the same as that of those which have been already described. When the speed decreases, the tension of the spring, at the end of the weight lever, overcomes the centrifugal effort of the ball, and the latter is forced in toward the shaft, carrying with it the end of the eccentric lever, and thus giving the valve greater throw, and extending the period through which the steam follows the piston, producing more power and bringing the engine up to speed. The reverse change of speed of engine produces the opposite action of eccentric and of valve motion, and the cut-off is shortened, and the power of the engine is reduced to that needed to give the correct speed. As this governor may be made as nearly isochronal as may be desired, the approximation to correct speed may be made as close as is consistent with the sensitiveness considered permissible. The use of a single eccentric and of a single governor ball, and the general simplicity of this combination, are especially pleasing to the engineer. They, however, include the use of a single valve, and thus restrict the designer, somewhat, in his adjustment of the steam distribution, a restriction which the more complicated forms of valve-gear are constructed to avoid, as it is well understood that the economy of the machine, in its use of steam, is to a certain extent, dependent upon the method of distribution of the steam entering, and of the exhaust leaving the cylinder. The main objection is the fact that the mean pressure of the steam entering the cylinder up to the point of cut-off is necessarily less with a single valve than with the gear introduced by Sickles and Corliss, and their successors, and which have been long standard, and which are admittedly superior in this respect. Whether the more costly, but more efficient gear shall be used, is to be determined partly by the cost of fuel, and must be settled by the judgment of an experienced engineer in each individual case.

The difference in this respect is not, however, as great as has been by some engineers supposed, and the economical value of heavy compression is now becoming so well understood, that the general impression in regard to this system of valve motion is becoming considerably and rapidly modified. What is lost by the drop of pressure between the boiler and the piston, is partly compensated by the variable and automatically adjusted compression obtained with this kind of motion, as is well illustrated in the action of the Stephenson link as used on the locomotive. With this arrangement, there is also some loss at the exhaust period, but not usually enough to be considered serious. As this

•

Belt

Shaft

particular engine is operated, this latter loss, and possibly, to a slight extent, the former, are somewhat reduced by setting the valve without lead, or even with "negative lead," *i. e.* so that the engine does not take steam until the crank has just passed the center, and the piston is starting on the forward stroke.

The engine, as a whole, with all its important parts in section, is shown in the above engraving. The unusual quantity of material as compared with earlier practice in older forms of engine, the excellent distribution of that material, the small number of parts, the heavy crosshead, the arrangement of fly wheels, and the form of valve are all plainly seen, as well as the general arrangement and system of connection. The rods and pins, and all running parts, are made of steel ; journals are ground to perfect form and polished, and the engine, when completed, is set up in the shop and carefully tried before sending it out, as is becoming the custom with good builders everywhere. The designer has made a special effort to reduce friction to a minimum, and has given the engine easy running piston and crosshead, perfectly formed journals, and a valve gear and governor, which are as nearly frictionless as those parts can well be made. The growth of the engine into its present shape, from the first crude sketches made in 1869, to the finished engine and completed type of to-day, and especially the gradual evolvement of the governor and valve gear from the older forms, would be an interesting subject of study, but it cannot here be undertaken. The survival of the fittest, among these devices, has led to the production of the engine above described.

The Straight Line Engine has been frequently applied

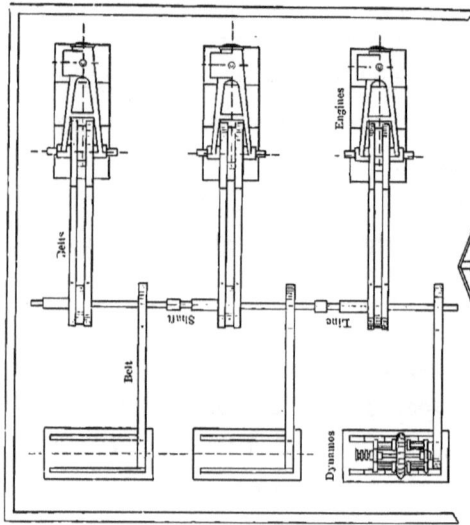

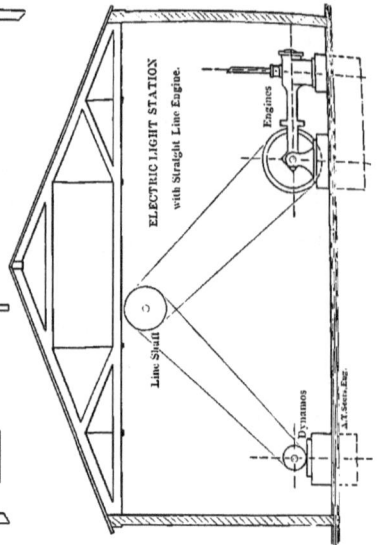

ELECTRIC LIGHT STATION
with Straight Line Engine.

to the driving of electric lighting apparatus. In Penney's arrangement of a station of 120 lights, the connection of power to dynamo is effected through friction clutches, which may, at any instant, be thrown out or thrown in ; any two of the engines have ample power to drive all three of the dynamos used, and a reserve is thus supplied to be used in case of the necessity of throwing off one engine for repairs. The current from any one generator is capable of being switched into any circuit, and all parts are accessible for examination and repair. A novel device is that of placing the driving pulleys, on the main line, on separate hollow shafts, independently supported, to prevent the springing of the line shaft by the pull of the main belts. The line shaft runs directly through the jack shaft, carrying the driving pulley on the line.

As this engine is adjusted, with large compression when at regular speed doing the rated work, with negative lead on the valve at that point, becoming positive lead at ¾ cut-off, it illustrates well the efficiency of the class. A 50 horse-power engine, driving a 40 light dynamo, according to the report of the manager at the station, ran at 219 revolutions, and at 220 when 27 lights were thrown off. The writer, testing one of these engines rated at thirty-five horse-power, using a Prony brake to take up the power, counted 233 revolutions, light, and 232, loaded with above forty horse-power ; with lower steam, the figures became 231 and 230. A well-balanced valve and a nearly frictionless governor are the elements giving success here. Every good engine, driving dynamos, is expected to rival this, doubtless, but, doubtless many do not. The single-valve engine can evidently, as is here seen, be made, by a skilful engineer, to do excellent work.

VERTICAL STRAIGHT LINE ENGINE.

IV.

Engines Capable of Direct Connection.—(*Continued.*)

THE engine last described, and that to be here examined, are the result of an attempt on the part of their designers to secure a form of engine which should not only be so proportioned and so arranged in the disposition of their details that they may be driven up to the speeds of rotation, now so frequently found desirable, without excessive jar, serious wear, or dangerous heating of journals, but which should also be so simple in plan, so inexpensive in construction, and so easy of repair, that the cost of maintenance, that great tax upon the proprietor of the average steam engine, should be reduced to the lowest possible figure.

In these engines, the possibilities in the direction of increasing speeds, are sought to be made the most of. Their market is not only to be found in the domain of the electrical generation of light, and electrical transmission of power, but in older fields of work as well. The loss of power in the "jack-shafts," or "first motion shafts," of mills and workshops driven by the low-speed engines is an item of no inconsiderable amount in many cases. The tendency is now observable toward the adoption of the high-speed engine, even where not quite as economical in the use of steam, in direct connection with the main line of shafting, through the intermediary of a single belt or pair of gears, or even by directly attaching the crank-shaft of the engine

THE ARMINGTON & SIMS ENGINE.

to the main line by a coupling. Many flouring mills and several rolling mills to the personal knowledge of the author, have been operated in this way for some years, and the system will probably become rapidly more general. In this country, the use of gearing for such connections has long been almost entirely superseded by the introduction of belting. The smaller first cost, the diminished noise, the lessened danger which accompanies their failure, and other obvious advantages, have been found to far more than counterbalance the cost of maintenance of the belt. By thus connecting directly to the main line, also, the cost of belting is greatly reduced. As the speed of shafting is rarely less than 150, and seldom more than 250, revolutions per minute, it is not difficult or objectionable to establish this method of connection. The same advantages are then derived that are experienced in the direct connection of the engine to the dynamo-electric machine. The total first cost of power is thus often reduced thirty and sometimes 50 per cent. As has been already intimated, there seems to be no nearly reached natural limit to the increase of engine speeds, except the practical limit of perfection of workmanship and excellence of materials, which limit is being constantly pushed farther and farther back, as the demands upon our engineers and mechanics are more and more exacting. President Westmacott, of the British In-stitution of Mechanical Engineers, has remarked that, at the high speeds (400 to 500 revolutions per minute) attained by the screws of Thorneycroft's torpedo boats, the engines seemed to run more smoothly than at lower speeds. This has been noted by every builder, and every driver of fast engines. The author, in handling naval screw engines of short stroke and high speed, has frequently observed this fact, and, after a somewhat wide range of experience with engines of long and of short stroke, of from 15 to 500 revolutions, and of powers ranging from the toy engine built during his hours of leisure when a boy in a short jacket, to marine engines rated at above 5,000 horse-power, at sea and on shore, in the mill and the workshop or on the locomotive, he has never yet seen evidence pointing to any as yet nearly reached limit to engine speed, except that which is imposed by such conditions as we are gradually and steadily modifying, as our knowledge and skill become more nearly able to cope with the difficulties which arise in our constantly changing practice.

It will have been observed that, in all the engines which have been here described as adapted to direct connection to the dynamo and to the "first motion" shaft, some form of balanced valve has been used. It has been seen that one of the conditions of good regulation by a governor, which determines the "point of cut-off," is that the work thrown upon the governor shall be the least possible. This condition evidently points to the use of some expedient, in cases in which a positive-motion gear is used, by which the resistance to motion of the valve, while a change is being effected by the governor, shall be made a minimum ; this evidently indicates the advisability of adopting some form of balancing device.

The engine to be here described has been designed with this end in view, as well as with the idea of securing a form of machine which should be simple and inexpensive to build, and to keep in repair ; prompt and exact in regulation under sudden variations of load, and as nearly isochronous in its governor-motion, as is practicable. It is of the same general class with the last several described forms

GOVERNOR AND ECCENTRICS.—MINIMUM THROW.

of engines, but differs from them in its details and in its proportions, somewhat, and, especially, in the form of its valve, and in the devices intermediate between governor and valve. In this engine, the "piston" valve is used, combined with a double port, such as was first used by Allen in the locomotive slide valve. These details are illustrated further on. The engine, as a whole, will be first described.

The accompanying engraving present two perspective
views of the Armington & Sims Engine, of the styles commonly used in driving electric light machinery.   The bed
is seen to be of the kind already described in the account
of the Porter-Allen engine, heavy, solid, stiff, yet neat, and
even graceful, taking the bending stresses of the guides at
its upper surface, and insured against twisting strains by the
box form of its section.   Two main pillow blocks, in the
first engine illustrated, carry its steel crank-shaft, and support the two wheels, one of which is a balance wheel, and

CRANK-PIN AND " WIPER."

the other of which is the pulley, from which the engine is
belted to its work.   The steam cylinder is overhung, and
the exhaust pipe is carried down below the floor, clear of
the foundation, which latter has a minimum extent, and
cost, while amply heavy, and is long and strong enough to
carry the engine steadily.   In some cases, the frame is made
with but one pillow block, and the crank is overhung; the
plan here illustrated is, however, a better one when the engine is to be driven up to the now usual speeds of such
machines.

The journals are all large, and carefully calculated for
the speeds and pressures adopted.   The designers make use

THE ARMINGTON & SIMS ENGINE.

of a method of calculation introduced some years ago, by the author, which is based on the working of marine and stationary engines, under his own management, or under his own observation. The drain-pipes for the cylinder are fitted as usual, but should be rather larger and more carefully planned, than is necessary where the engine has a valve, which may lift from its seat should the boiler at any time " prime" or "foam," and send water over into the cylinder with the steam. The provision for lubrication is a matter of vital importance in all engines of this class. In this engine the "sight feed " is used, in which each drop of oil falls through a clear space, on its way to the point to be oiled, in full view of the man in charge, and any failure of the oil to "feed " is thus promptly detected. The crank-pin is supplied by a "wiper" (see Fig.), which takes its supply of the lubricant from the oil-cup at every revolution of the crank. This device has been used, in very similar form, by the author, on fast marine engines, with perfect satisfaction, and it is found to work well here.

The two large engravings show opposite sides of the engine, and the second exhibits the arrangement of a single wheel, and of the steam-chest and valve mechanism. As is here seen, a governor, of the same type as that exhibited in the articles describing the "Buckeye" and the "Straight Line" engines, is secured to the arms of the pulley on the nearer side of the frame, and is arranged to adjust the position of the eccentrics, which give motion to the valve through a rod and valve stem, the connection between which two parts is made at a point at which they can be conveniently supported by a rockshaft and arm carried at

the middle of the length of the frame. The cranks are, as
seen in both illustrations, two discs in which the balancing
mass can be secured at any desired point. The width of
the pulley carrying the main belt is sufficient to take a belt
of such breadth, that the stress shall be about 35 pounds
per inch of its width. The main bearings are made with
boxes set at an inclination to the horizontal, and provision

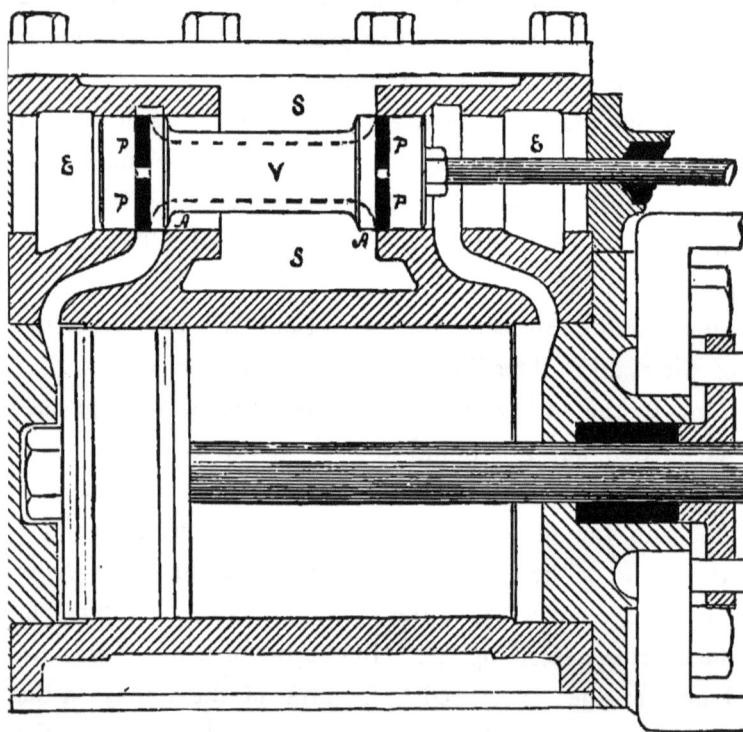

SECTION OF CYLINDER.

is made for taking up wear. The crank-pin is of steel,
ground carefully to size, as is the universal practice among
good builders of this class of engines. In this machine the
main journals are also ground. The distance between main

bearings is made as small as possible, to permit high speed with little risk of springing the shaft. The front cylinder head can be removed, when necessary, as shown in the next illustration, independent of bed and cylinder alike.

As here shown in section, it is seen that the cylinder, steam-chest and valve-seat are all in one casting, which is, however, not a remarkably intricate one. It is best shown by the perspective view, while the section next given will afford a better idea of the arrangement of the valve.

The steam-chest, S, S, is in direct communication with the boiler, and the valve, which is of the piston form with a double steam-port (the second port being seen at P, P), is surrounded by the "live steam," thus taking steam at the middle and exhausting it at the ends of the chest, at E, E. The valve moves precisely as does the ordinary locomotive slide valve, and, as here shown, is just taking steam at the piston end of the cylinder, both directly past the shoulder of the valve and through the secondary port at the opposite end of the valve. Thus the steam is introduced, at the beginning of the stroke, through a double length of port, and hence, with unusual promptness when the engine is running at high speed. The consequence is that it gives approximately boiler pressure in the cylinder, and throughout the stroke up to the point of cut-off, if the steam pipe is short and direct, the steam line on the indicator diagram is very nearly perfectly horizontal and straight from end to end. This is a very unusual feature in diagrams from engines having positive-motion valve-gear. The form of this valve is well shown in the accompanying engraving, which exhibits the valve apart from its casing.

All engines of this class will have been seen to be remarkable for the shortness of their stroke of piston, as
compared with the diameter of cylinder. The section of
the cylinder just given, shows how advantageous is this
proportion in enabling the port-space to be reduced to a
comparatively small volume. In the engine of long stroke,
the port-space becomes seriously large and the compression
required to fill it introduces a considerable loss both of
power and efficiency, if the valve-gear used is of the type
here seen. In fact, it would be probably quite impracticable to secure such a steam distribution as would satisfy
the majority of engineers, were the engine of long stroke
and a single valve adopted moved by a link, or by such an
equivalent for the link as is here used. The total "dead
space" in these engines, including piston-clearance, is
sometimes as low as 5 per cent. on large sizes. In all cases
compression fills this space at every stroke. The piston-
valve has been often used by earlier builders, but that here
shown possesses a novelty in the double port. Its advantages are the ease and cheapness with which it can be made
and fitted, and with which it can be replaced when worn, its
perfect balance and ease of working under any practicable
steam pressure, its permanence, tightness and remarkable
durability when properly cared for and used with boilers
supplied with good water. Its disadvantages are, the ra-
pidity with which it sometimes wears, when it is not kept
well lubricated, or when it is exposed to the action of steam
carrying over from the boiler acidulated or dirty water, the
danger of injury to the cylinder or its heads when priming
occurs, and the proneness of the attendant to neglect its

repair when it requires such care These disadvantages have sometimes proved to be so serious, as to give many engineers a very strong prejudice against the valve ; on the other hand, this unfavorable prejudice seems to be now giving place to a decidedly favorable opinion, assuming that the valve is well made and is to go into good hands, and to be used under proper conditions, and these and some other very successful makers have definitely adopted the piston valve as a feature of their standard designs ; it is even coming into use in marine engines of the largest size. In

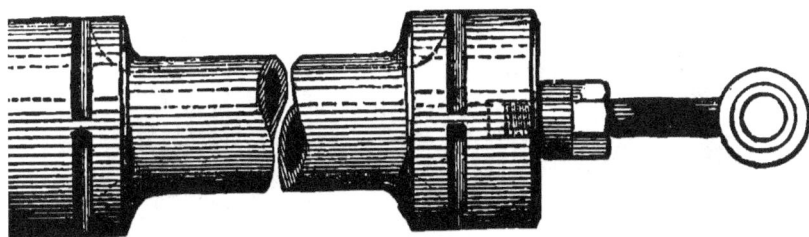

ARMINGTON & SIMS VALVE.

the engine here under consideration, the valve is said by the constructors to have proved eminently successful and to have proven more durable than their earlier constructions, in which they adopted a balance flat valve. It is probably too early, as yet, to fully decide what are the exact relative merits of the two kinds of valve. In this particular case, the removal and replacement of the piston valve can be done quickly and inexpensively, and a spare valve being kept on hand, it is probable that its use may prove economical and satisfactory even where the water used for the boiler is not of the best.

One of the most important, novel, and beautifully ingenious details of this engine, is its peculiar arrangement of governor and eccentrics. These parts are exhibited in two engravings.

The regulator is precisely the same, in principle, as those already described as adapted to the adjustment of the eccentric on the main or the governor shaft. It has the two weights, 1, 1, carried on, and forming a part of arms piv-

Armington & Sims Governor and Eccentrics.—Maximum Throw

oted to the governor pulley, and revolving in the vertical plane as usual in that class of governors. The position of these weights, as determined by the speed and the action of the springs, determines the position of the eccentrics, C, D, and thus the position and motion of the valve, and the point of cut-off, flying out and giving a higher ratio of expansion as the load on the engine is diminished, or as steam pressure rises in the slightest degree, and a lower ratio as these conditions are reversed. In the device here adopted, however, the valve is driven by an eccentric which is "duplex." One eccentric, C, is set inside another, D, and connected to the governor arms in such a way that, as the weights separate with increasing speed of engine, both eccentrics are turned on the shaft so as to cause their "throws" to coincide, or to separate, as may be necessary. When they coincide, the travel of the valve is due to a greater total throw, B, and is a maximum ; when they are separated as far as possible, the throw becomes A, and the travel is reduced to a minimum. The action is almost precisely the same as that of a "Stephenson-link," worked between full and mid-gear. When the two eccentrics give maximum travel, the action is that of the link-motion in full gear; when they are at opposite sides of the shaft, the action is that of a link in mid-gear. By setting them at intermediate points, the throw is made that is required to give an intermediate action of the valve, and thus the distribution of steam is made to accord with the demands of the work by such a variation of the ratios of expansion and of compression as is obtained by the link-motion, and, in this case, with the advantage in promptness of opening and of closure obtainable with a double-ported valve. The range of action given in this engine is sufficient to permit a

range of cut-off from o to about three-quarters stroke.   The
lead remains unchanged, and the compression increases as
the ratio of expansion is increased.

The springs of the governor are used in compression.
The distribution of steam at the usual speed, and with full
load, is shown by the accompanying illustration, which is a
copy of an indicator diagram taken from one of the engines
driving the large dynamos at the Edison station in New
York city.   These engines are coupled directly to armatures,

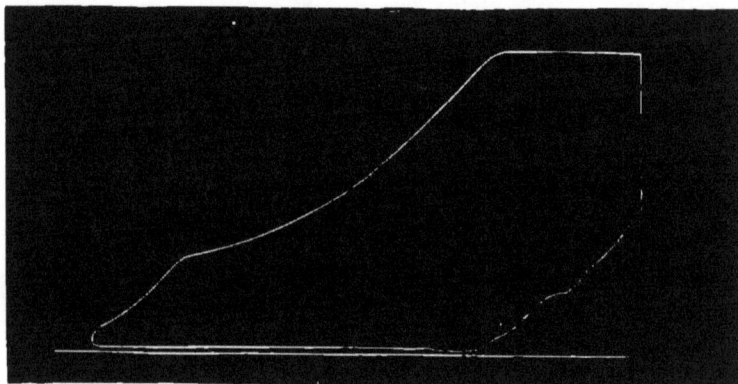

DIAGRAM TAKEN AT THE EDISON STATION.

and make with them 350 revolutions per minute.   One of
these engines was recently kept at work 17 days, making
over 8,400,000 revolutions without stopping, and then was
not stopped because of any difficulty with the engine.
When examined by the author, they were doing their work
steadily and smoothly, and were not appreciably affected
by the sudden changes of load produced by throwing on
and off any considerable proportion of the lights on the
circuit.

This engine illustrates well the perfection of regulation attainable by these positive motion valve-gears attached to this form of governor, to which attention has already been called. At a trial of engines of this make made by the author, to satisfy himself in regard to their action under varying load, 25, 50, and sometimes 60 Thomson-Houston arc lights were thrown on or off, and the variation of speed was but one and two revolutions, respectively, in 280. No special preparation or adjustment was allowed in this case, and there is no reason to doubt that still closer regulation and more perfect isochronism are attainable, if they, at any future time, should prove to be desirable. These engines, 9½ by 12 inch cylinders, had never been before tested, and had done no work until started under the direction of the author. The lamps demanded very exactly 0.7 horse-power each, a fact which indicates that, as connection is there made, there can be but little lost power between the engine and the lamp. The form of card under load is seen below.

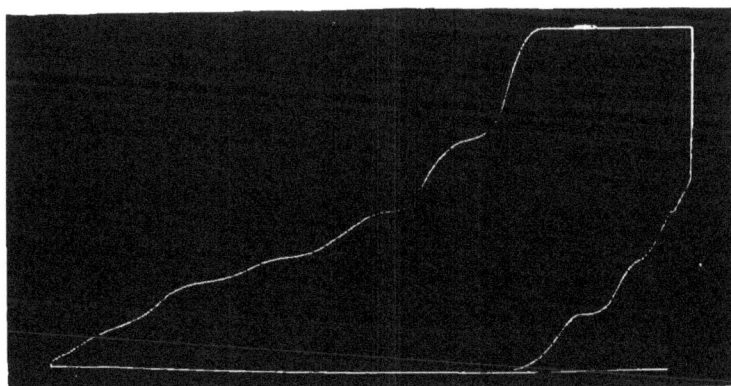

DIAGRAM TAKEN BY THURSTON.

The success here obtained in the use of a single valve is as encouraging as it is remarkable. While it can hardly be expected that the economy of this system, other things being equal, can be fully up to that obtainable with the more elaborate forms of valve-gear previously illustrated, there is no question that it is so great that these simple forms of engines will be able to find a market in that very wide field in which their extreme simplicity of mechanism and their moderate cost, as well as their successful operation at high speeds, are qualities which compensate any such differences in cost of the steam supply. If the same distribution of steam, and the same economy is obtained with the one form of valve motion as with the other, and if, as is the case to a very satisfactory degree with these engines, a correct form of indicator diagram can be obtained, it is to be expected that the engine will be economical in its use of steam. The increasing compression here noted with increasing expansion is a decidedly advantageous feature, as it has an important influence in checking losses by "cylinder condensation" at high ratios of expansion, while also reducing the waste due to large clearance spaces, where such exist.

Every engine and every machine of importance, or remarkable in any respect, as in such a combination, of ingenious devices, effective combination, and efficient operation as is here illustrated, is, invariably, the outcome of a long period of progressive invention, unintermitted experiment and more or less steady growth from an initial stage to its condition of successful adaptation to the demands which it is especially fitted to meet. The Armington &

Sims engine is no exception to the rule, and its inventors and makers, as has been seen, are fortunate in having been able to reap so satisfactory a harvest after so long a period of growth and ripening. The engine is now built, not only in the United States, but in Canada, Great Britain, France and Austria. This American engine is in use on many foreign steamers, and in numbers of European buildings, public and private. It drives the dynamos in the British Houses of Parliament.

V.

Fast Engines of Peculiar Design.

THE forms of steam engine which have been described in the preceding articles have been chosen as being fairly representative of what may be termed standard types of engine as built by makers of reputation. It will be seen that they present to the student of the steam engine several distinct forms of machine, each of which is now acknowledged to be well adapted to produce a certain result in the application of heat energy, through the medium of steam, to the production of power, and that each is especially fitted to do its work under certain definite conditions, which conditions are less completely met by the others. Each is well-known in the market as an engine which has taken its place among those which have passed the experimental stage and may be relied upon to do good work if well built and put in operation under the conditions that it is designed to meet. They embody ideas and inventions which have grown into form during years of experiment and faithful trial and the variety of makes to be found in the market belonging to each class, and differing only in the design and construction of details, proves that the main principles upon which each class is based are well established and sound.

The engines now to be examined are distinguished by certain peculiarities of design and construction which mark, in some cases, new departures, in other cases, peculiar ways of reaching the end at which more familiar devices were aimed.

It has been seen that the regulation of the steam engine has been found to be one of the most important matters to which the attention of the engineer has been called. For many purposes, the uniformity of motion of the engine is an even more important quality than its economy in the use

THE BALL ENGINE.

of fuel, or in all running expenses. A slight change of speed in an engine driving dynamo-electric machine will seriously injure the value of the light, in nearly every location, and may sometimes entirely destroy it ; a moderate variation of speed in the motor of a cotton mill making fine goods may break more threads in the spinning department or do more injury in the weaving room, than would be compensated by the difference in economy between the most efficient " automatic" engine ever made and the most wasteful engine in the market. The principle of regulation of the steam engine has been, from the time of the application of the old "fly-ball" governor to the Watt engines of a century ago to the present day, that of making the speed of the engine determine the amount of steam that shall be supplied to it. In the first engines used in the driving of machinery, in the old "Albion Mills" erected by Watt and his partners in London, in 1786, and for 50 years afterwards, the governor adjusted the supply of steam by moving a throttle valve. The governor was next arranged to determine the point of cut-off by Zachariah Allen, of Providence, . R. I., in 1834, and by George H. Corliss, in 1849, to adjust the trip of his detachable valve-gear. From this latter date, it has been the universal custom to so apply it in all engines in which uniformity of motion and economy in the expenditure of steam were the controlling considerations in their design. The method of accomplishment of this result has been seen in the preceding pages, as practiced by Corliss and Greene, and by the constructors of positive-motion gears which have been the later outgrowth of modern changes in the application of steam power.

Now, after half a century since the grand step taken by Zachariah Allen has passed, and a generation after that taken by Corliss, a new principle has been introduced into the construction of the steam engine, viz., the control of the speed of the machine, so far as it is due to the varying load, *by that variation of load*, making the cause of the irregularity of motion its own corrective, and placing the regulating principle between the work and the engine in such a way that the latter may be made to preserve any given speed with perfect uniformity, so far as it depends on the load, or causing the speed either to be increased or diminished to any desired extent by any given variation of load.

This idea, like all valuable inventions, has not been the result of a single thought or the product of a single brain; it has been floating in the minds of thoughtful engineers for a long time. It was proposed to the author, by one of the generation of inventors just passed away, years ago; but, in its present form, it became practicable only after the introduction of the high-speed engine had permitted the use of the form of centrifugal governor seen in the engines last described. The engine about to be considered embodies the first practically useful application of this principle, in a practically successful form of engine.

The Ball Automatic Expansion Engine is the invention, so far as it differs essentially from other engines of its class, of Mr. F. H. Ball, of Erie, Pennsylvania. In its general form and in the details of construction, generally, it resembles the last two engines which have been described. It has a single-valve, positive motion valve-gear, and the solid compact structure characteristic of all the so-called high-speed engines. The accompanying illustration will give a correct idea of its form and proportions.

The engine bed is of strong and stiff construction, and very similar to others with which the reader has become familiar. The steam-cylinder is overhung and bolted to a faced flange as in the Porter-Allen engine. The main pil-

low blocks are set in the bed of which they form a part, and
their caps are placed at an angle with the horizontal plane,
as is sometimes done in marine engines, and less frequently
in stationary engines. The system of boring the seat for
the cylinder, aligning the guides for the cross-head, and
boring out shaft-bearings, here adopted, gives perfect alignment ; and the preservation of the alignment is insured by
this unification of parts formerly detached. As is the case
with all good engines, the fitting parts are made to standard
gauge, and a system of inspection insures good work. Packing is dispensed with, and joints are made tight, by securing
exactly plane, and perfectly smooth, surfaces, at abutting
points. The wearing surfaces of the valves, and other
rubbing parts, are scraped to shape and exactness of form,
by the aid of surface plates. The valve is made tight under steam-pressure, the form of the valve being such as to
permit this rather unusual operation.

The Ball Engine has a short stroke and high speed of
rotation, ranging as now built, from 7 to 10 inches diameter
of cylinder, 10 to 12 inches stroke of piston, and making
250 to 350 revolutions per minute.   These proportions are
adopted, probably, principally with a view to meeting the
demands of electric lighting.

The essential and most peculiar feature of the Ball engine, and that which gives it a place in this little treatise,
is, as has been already stated, its governor.

The Ball Governor is, in the main, like the governors
which have been described as controlling the several engines
which have been immediately herein before described.   It
consists of a "governor-pulley," from the arms of which

are swung a set of weights, which are arranged to move in
the plane transverse to the shaft on which the pulley is carried.   These weights, or balls, are restrained from moving
outwards, under the influence of centrifugal force, by a set of
strong steel helical springs, secured, at one end, to the balls,
and at the other, to the rim of the pulley.   Any movement
of the weights, in either direction, causes a motion of the

THE BALL GOVERNOR.

eccentric, resulting in the alteration of the throw of the
valve in such a direction, and to such an extent as to bring
the engine very exactly to speed.   To this extent, the Ball
governor is identical, in its general construction and in its
principles and mode of action, with those already familiar
to the reader.   To this extent, it is possessed of the same

qualities as the others of its class, and it has been seen that good workmanship and correct proportions and adjustment may give wonderful nicety of regulation.

To this governor, as commonly built, Mr. Ball adds a remarkably ingenious, and singularly simple yet perfect, invention ; it is exhibited in the accompanying figures. The first of these illustrations shows the governor-pulley detached from its shaft, and does not show the eccentric; it presents only the essentially novel part of the device.

It is seen that, attached to the radius-bar of each ball, is a small spring, connecting a point near the fulcrum of that lever with the extremity of a strong, peculiarly shaped arm, projecting from the hub on the shaft which is seen within the hub of the pulley. The governor-pulley is set loosely on this inner hub, which latter is keyed fast to the shaft. The arrangement is evidently such that, the shaft being turned by the engine, the effort must be transmitted through the small spring to the weight arms, thence to the pulley, and from the latter to the load to be driven, through a belt carried on that pulley. The effect of this curious disposition of parts is easily seen : Suppose the governor to be so adjusted that, at normal speed and under the rated load, the supply of steam and the distribution of that steam, are precisely correct, as intended by the designer of the engine. Now, if a variation of steam-pressure should occur, the governor at once meets the consequent change of speed by a corresponding change of steam-distribution, and the variation of speed is restricted to a range, which, if the governor is well proportioned and well adjusted, may be quite imperceptible to the senses, and hardly measurable by count.

This governor here acts like all the others. But, suppose the steam pressure to be unchanged, and the load to vary—we now have a new movement introduced. The force exerted in driving the load is transmitted through the small springs which are peculiar to this governor, and which connect the main shaft to the driving pulley, through the governor. The instant that any relaxation, or any increased tension, is felt here, the relaxation or the extension of the springs, so produced, causes a change in the position of the weight-arms, and a corresponding alteration in the position of the eccentric ; and the steam supply is at once readjusted to meet the variation of load. This may be done so promptly and so exactly, that, however much the load may vary, the speed of the engine remains precisely the same. Load may be thrown on and thrown off to any extent that may be found desirable or necessary, and the engine goes on with its fluctuating task without an instant of visible change. Should both steam-pressure and load vary at the same time, the load.strings set the example of changing the steam distribution to meet the new conditions, and the governor-springs controlling the balls are immediately seen to yield to the effect of the varying steam-pressure, and to continue their motion until the flying weights have set the eccentric in correct adjustment to give the right speed. If the governor is perfectly isochronous, the new adjustment meets the case exactly, and the engine runs at the intended speed as before. The load-springs may even be so adjust-ed that an increase of load may produce a decrease of speed to any desired extent, or, even more commonly and usefully, so that *an added load may give increased speed.* This latter is done in some cases when driving electric lights, and also in saw-mills, and for other kinds of variable work. In the former case, the engine is adjusted to give standard speed when driving full load, and to reduce its speed as lights are turned off ; in the latter, the engine runs at speed while the saw is cutting, and slows down when the work is off.

The next figure shows the eccentric. *A* is the main eccentric having an elongated shaft opening ; to this eccentric is attached the arm *B*, of which the outer end is pivoted, allowing the eccentric to swing across the shaft; this motion controls the time during which steam is admitted, each stroke. This swinging motion is controlled by the rotation of the disc, *C*, in the following manner : The disc has a flange, *D*, on its side, which is eccentric to the shaft, and on the inside of this eccentric flange is a ring, *E*, which engages with a stud, *F*, in the main eccentric. Thus the rotation of this disc forward and backward causes the eccentric to swing across the shaft. The disc has a sleeve encircling the shaft and projecting through the elongated shaft opening in the main eccentric, and on the end of the sleeve is a flange nut, *G*, which holds the parts in place. The rotation of the disc is produced and controlled by the governing forces ; the centrifugal force of the weights met by suitable springs ; and the resistance of the load equilibriated by the centrifugal force of the weights.

This form of governor is a very safe one, as, should breakage of load-springs occur, the engine slows down or stops. The risk of injury of this kind is unimportant, however, if the springs are properly made, as the load carried by them is insignificant. A 50 horse-power engine, at 300 revolutions per minute, carries a load of but about 500 pounds on each load-spring. If correctly proportionated and made, they should endure indefinitely. The endurance of all these springs is the greater for the periods of rest frequently given them, and for the fact that they are, much of the time, under very uniform tension.

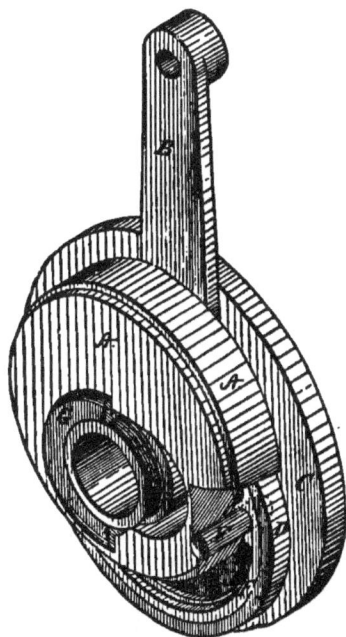

THE BALL ECCENTRIC AND CONNECTIONS.

The practical result of this novel modification of old methods of regulating the engine is that the regulation of the steam-engine now can be made to cover more than the simple preservation of a fixed velocity of rotation.   It is now possible to determine, within certain limits, not only what degree of variation from normal speed shall be permitted, but also what shall be the normal, and if desired, varying, speed of the machine, with varying load.   It may not only be made to run at a certain fixed speed, but may be caused either to increase or diminish the speed, according to a fixed, and economically desirable, law.   This new principle will probably find many applications, although such problems have rarely come to the consideration of the designing engineer, hitherto.

The accompanying peculiar diagrams are taken from the recording apparatus of the "Moscrop Indicator," an instrument which automatically and continuously records the speed of the engine and its variations.   Each revolution produces a dot, the height of which above the base-line indicates the speed.   The first of the two diagrams is from an

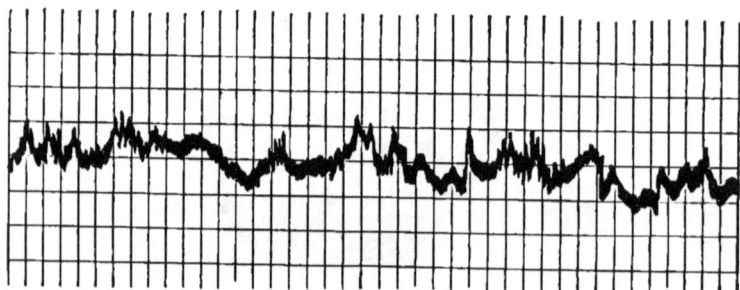

MOSCROP SPEED DIAGRAM.—FAIR REGULATION.

engine of 250 horse-power, fitted with an "automatic cut-off," and furnishing power to a paper mill.   It is claimed to do good work ; but the author has no personal knowledge of it.

The second is furnished, by the owners of the Ball Engine, as illustrating fairly an equally trying case. The author has other cards of this kind which, with great variation of steam-pressure, nevertheless are very smooth, although not as smooth as that here reproduced. They are also interesting as showing how useful a recording speed-indicator may be. Such records are more satisfactory, in comparing speeds of engines, than are even the best of counters, and vastly more satisfactory than counting by the watch, as they exhibit the rate of each revolution, together with the variation of rate for extended periods of time.

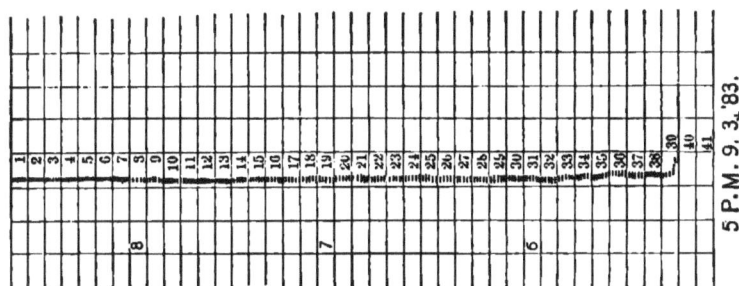

MOSCROP SPEED DIAGRAM.—BALL ENGINE.

This engine, with its novel governor, is one of the most interesting products of mechanical ingenuity that has been seen since the days of Watt. It will probably have but little influence on the vitally important matter of steam-engine efficiency, as that term is customarily applied, that is to say, upon the economy of the engine in consumption of steam and of fuel ; but it will undoubtedly, in many of its applications, be found to have a very important effect in adapting the engine to its work, and upon its efficiency in that relation.

THE IDE ENGINE.

THE engines which have been described are by no means the only engines which are deserving oi mention, and of careful study, as illustrating the peculiarities of the best modern practice in the field which it has been the object of the author to explore. A number of other engines, of one or another of the classes which have been described and illustrated in the preceding articles, have nearly or quite equal claims for consideration. Of these engines, only typical or representative examples have been sought, and have been selected from the machines with which the author is most familiar. One more engine may be here described—not as possessing the singular novelties of design which distinguish some of those already examined, but as affording a good illustration of the principles and practice which have come to be recognized as distinctive of the latest phase of that progress, which has recently been so rapid, in the direction of improved methods of construction, as well as of design, and in the application of the modern materials of construction. The engine is one with which the author cannot claim that personal familiarity which has led, in some cases, to the selection of those which have been previously considered; but a description, such as is to be here given, will show that it may fairly be taken as a representative of the best practice, in matters of detail, which it is the special object of the writer now to exhibit.

The Ide Engine is of the same class with all the engines described in the preceding section—a high-speed engine, intended to be driven up to high power and to occupy small compass; to regulate with all the accuracy desired in electric lighting, and in the spinning of fine cotton; to have good wearing qualities, and to be economical in its use of steam and of fuel. The illustrations exhibit its general form, and the more important details of the machine.

Examining it in some detail, it will be observed that the frame, although of novel design, is of the same general form with those which have been already described in this class,

possessing that solidity and rigidity that have been seen to
be an essential feature of all successful high-speed engines.
The main pillow-blocks are formed in the frame; and the
cylinder is secured at the opposite end, overhanging as in

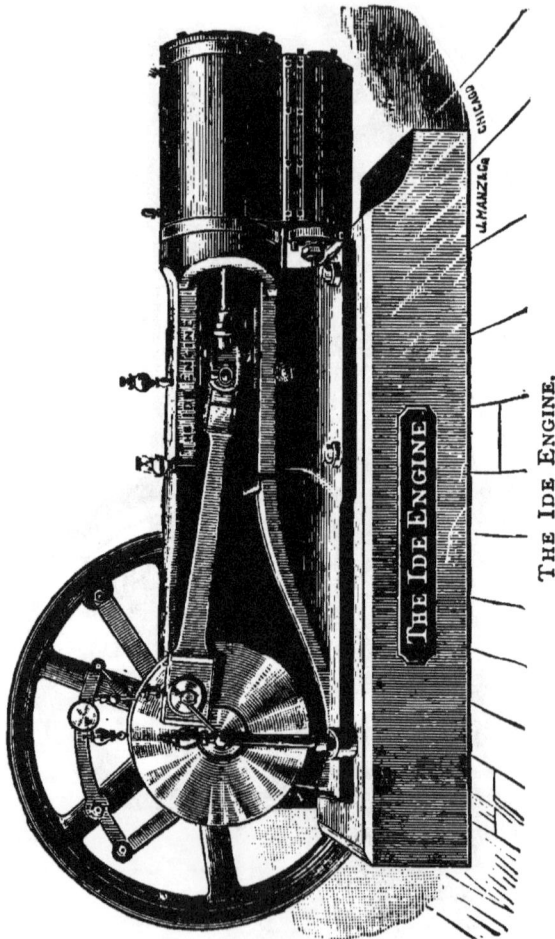

cases already familiar to the reader. The crank-pin is set
in a disc, which permits counterbalancing, and gives great
strength. The connecting-rod is tapered from the crank-

in to the crosshead-end, in the manner now common to all fast-running engines. The outlines of all visible parts indicate strength and stiffness, and are very neat in design.

The valve-gear and governing mechanism are shown best by the view of the opposite side of the engine, given in the next engraving. The piston-valve is adopted, and is placed directly under the steam-cylinder. This arrangement permits most complete drainage of the cylinder, and thus lessens the danger of accident, should the entrance of water with the steam occur to any serious extent. The placing of the valve at the side is not an unusual feature of this class of engine; but the arrangement here adopted is, in this respect, still more advantageous. This arrangement also affords a means of getting an equalization of the travel of the valve relatively to that of the piston, which is an advantage. Still another advantage is that this position of the valve-chest gives dry steam from the steam-chest, by causing it to act as a trap, as well as drains the cylinder of water that may have condensed within it. The connection with the steam pipe is made above the line of connection between the steam-chest and the cylinder, and it is thus rendered possible to remove the former, and get at the valve without disturbing the steam pipe.

The regulation is effected by a governor of the class adopted in all engines of this kind, and the regulation and the action of the valve are similar in character and in precision to those seen in engines already described. The range of power, and the distribution of steam at various points of cut-off, are shown very beautifully in the indicator diagram here given, which was obtained by suddenly throwing off the load; each revolution gives a distinct "card." Steam may follow from the beginning nearly to the end of stroke, with good exhaust and an excellent range of compression. The speed of engine was here 225. The card was taken by loading the engine to its maximum power by a Prony brake, and then taking the diagrams while the governor was adjusting the steam supply, the brake being at the moment released. The smallest card is therefore a "friction card." The smoothness of action of the regulating mechanism is shown by the uniformity with which the power falls off and the cards diminish in area.

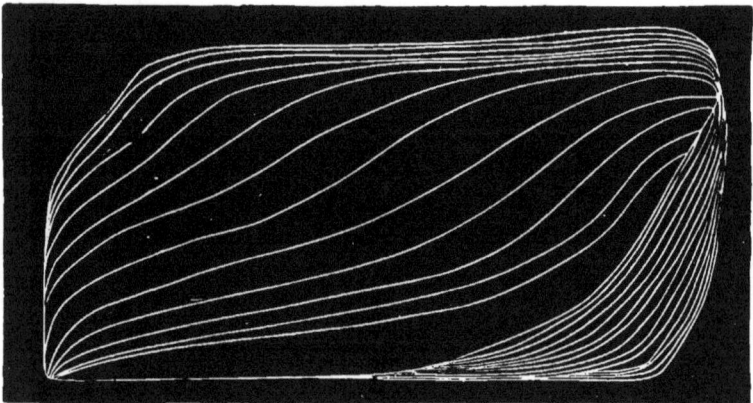

SERIES OF INDICATOR DIAGRAMS.—IDE ENGINE.

The next diagram shows the range of work which such engines are capable of doing, and illustrates very finely the change in the distribution of steam which takes place in this accommodation of the power of the engine to its load. It is seen that the compression, as well as the expansion, gradually changes in amount as the power varies, both acting to reduce the area of the diagram with diminishing power, or to increase it as the required power becomes greater. A very interesting effect of this change is to give increased economy in the use of steam by checking cylinder condensation, the greatest known source of waste of heat, just when that loss becomes most serious in both absolute

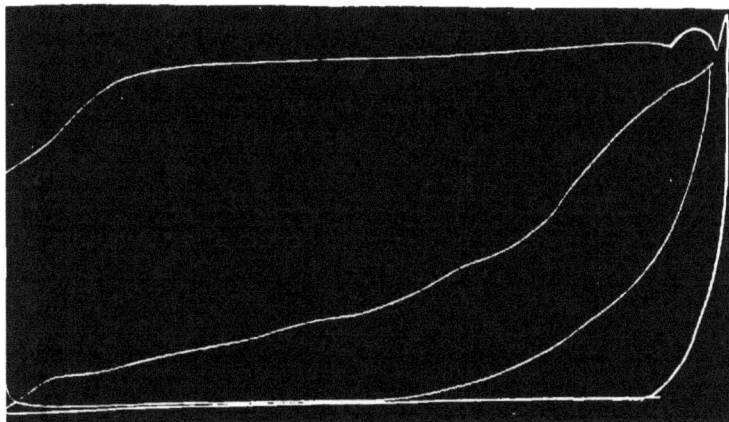

INDICATOR DIAGRAM.—IDE ENGINE.

and relative amount. In some cases, the economy obtained, with considerable expansion, by the introduction of large compression, has amounted to above 10 per cent. Where superheating is adopted, this gain is less; but in the usual case, using saturated steam, the use of the valve-motion, of which an example is here illustrated, brings with it a very important advantage; and nearly all builders of such engines are now agreed in testifying to its value. The lines of indicator diagrams obtained by the author from this engine are unexcelled by any that he has yet seen from engines of this class.

One very important feature of recent progress in the construction of the steam-engine is well illustrated in the Ide

Engine, and affords a special reason for studying it; this is the extensive use of steel in its running parts. Within a few years it has become possible to obtain from the makers of Bessemer and "Open Hearth," as well as of crucible, steel, a quality of metal which earlier could not have been obtained at all. This is a steel which is distinguished, chemically, by its low percentage of carbon and its relatively high proportion of manganese, and physically, by its wonderful combination of ductility and strength. As the proportion of carbon decreases in steel it loses strength; but it gains ductility and malleability in a far higher ratio, and thus it happens that the softer qualities are much better fitted for use in machinery than are the very best of wrought irons produced by the ordinary process of puddling. The former are strong, tough, amply hard for all such uses, and perfectly homogeneous; the latter are less tenacious, often not as ductile, and are never homogeneous; but are full of "cinder streaks," and have a fibrous structure that is objectionable, and is never seen in steels. These steels are all made by casting molten metal into ingot moulds, and thus securing comparative freedom from cinder and defective structure.

The soft steels are displacing iron in every direction; and the probabilities seem to be that in the course of time, in the coming "Age of Steel," iron, puddled as is now usual, will be entirely displaced by these, properly so-called, "Ingot Irons." The Ide Engine, as well as other engines now com'ng into market from the shops of the best builders, illustrate this change of material. It has its piston-rod, its connecting-rod, its valve-stems and links, and its smaller journals, all of steel. Large castings are not usually made in steel in this country, but all small parts are coming to be made in that remarkable metal.

ENGINES OF THE NEW YORK SAFETY STEAM-POWER CO.

In the course of the somewhat extended series of descriptions of standard forms of engine which is now soon to be closed, it will have been observed that the tendency has been toward the reduction in number of parts, and increasing simplicity of mechanism as the speed of engine is increased. The earlier types of engine having detachable cut-off apparatus as a part of the valve-motion were engines of moderate speed of piston and of comparatively long stroke, and, therefore, of even more moderate speed of rotation. The latter forms of standard engine are of simpler construction, and of higher speed of piston, and of much higher speed of rotation. This difference is not only due to the necessity of reducing the number of parts and securing greater positiveness of action in the valve-gear, but it is also due to the more general recognition of the fact that economy of steam and fuel consumption is but one of the economies to be studied in the use of steam as a motive power, and that the cost of securing great economy of steam and fuel may be such as to more than compensate the saving effected by such expenditure. This is especially true of small powers, and common experience has shown that it is seldom advisable to construct complicated valve-gears for such engines, as the cost rarely comes within the commercially economical limit. This principle has probably been carried too far; and the author has no doubt that engines of

the higher grade may be often found commercially economical for even very small powers.   The field for the simpler
class of small engine, nevertheless, is enormously extensive;
and the number annually built is very great.

But little attention, comparatively, has been paid to the
design and construction of small steam-engines until very
recently.   The engineer has been too often inclined to look
upon this as too small a matter to demand the thought and
the time that he has freely given to larger and more at-
tractive work.   It is now different, and some excellent
forms of small engines are to be found in the market.   It
is the intention of the author here to describe a single example of this class of machine, not as the only good engine
of the class, but as a type of this class.

The British builders of portable and agricultural engines
were the first to develop the art of steam-engine design and
construction in this department.   A dozen years ago, they
were building engines of as little as 20, or even 10, horse-
power, which demanded but 3 pounds, and even less, of
coal per horse-power per hour.   As early as 1867, they
reached the figure 4.13 pounds;* in 1870, it became 3.73,
and, in 1872, the Reading Iron Works built an engine of 20
horse-power which, on trial at Cardiff, required but 2½
pounds of picked coal per hour and per horse-power.   This
engine had a cut-off valve on the back of the main valve.†
Single valve engines have never done as well; but some of
them have nearly approached these figures.   A consumption of 5 pounds of coal per hour and per horse-power is a

---

* Mechanical Engineering at Vienna; Reports on the Vienna Exhibition: R.
H. Thurston, Washington, 1878.
 † Ibid., page 100.

good figure, and is rarely attained in such small engines. The best of them may be expected to use from 5 to 7 pounds and to consume, therefore, from 40 to 60 pounds of steam, averaging perhaps about 50, on the basis of the indicated power.

Among the earliest of American engineers to turn attention to this department of mechanical engineering, were Messrs. Babcock & Wilcox, who have become well known as the inventors of a successful form of "sectional" steam boiler. The style of engine which was designed and introduced by them, and built by the New York Safety Steam Power Co., has now become as generally accepted as standard among builders of small engines as has the Corliss engine among constructors of drop cut-off engines. It has been copied in all parts of Europe, as well as in the United States. This may be taken as representative of the best methods of construction in this country, and as exhibiting the elegance in proportions, and that excellence of material and workmanship, which are now becoming recognized as desirable in steam-engines of even the smallest size. In fact, as has been seen, the opportunity here offered for improvement, and for economizing steam and fuel consumption, is much greater than with large engines; and these excellencies are, therefore, the more desirable.

The engraving exhibits the form of the engine here to be described. It is a "vertical engine" mounted upon a base-plate of neat and strong form, and with the steam-cylinder bolted by the lower head to a very strong and very graceful frame. The main journals are carried in bearing constructed in the frame, and consequently free from liability to loss of perfect alignment, or to unequal wear. The valve is either a plain, locomotive-slide, or, preferably, a piston valve. The latter is fitted in a detachable seat, which can be easily removed for renewal of seat and valve, should accident or wear ever make it necessary.

N. Y. SAFETY STEAM POWER CO.'S ENGINE.—5 H. P.

The vertical position of the engine prevents wear within the cylinder becoming serious or unsymmetrical. The pistons are hollow, and are packed with rings set with sufficient spring to keep them up to a bearing. The cross-head, which is shown in the following engraving, has its gibs turned to fit the guides in the frame, which latter are part of the casting of the frame and are bored out in line with the cylinder, and cannot possibly get out of line.

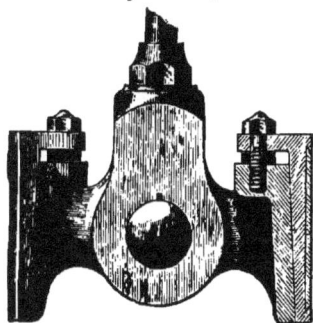

CROSSHEAD.

CROSSHEAD.

The engine above illustrated is of small size—4 or 5 horse-power—and has been especially designed for electric lighting purposes. The governor is that known as the "Waters Governor;" it regulates by adjusting the supply of steam passing to the engine through a throttle valve—a method which seems to have been here more successful than is usual in engines having to perform so exacting a kind of work. The speed of this engine is usually about 250 revolutions per minute.

Larger engines of this style are often constructed ranging up to 100 horse-power. The heavy engines, when of 15 to 100 horse-power, are given an independent crank-shaft pillow-block and a counterbalanced disc-crank. In these engines, of all sizes, the modern innovation of the use of steel for running parts is very generally introduced. The rods, pins, and minor parts are of this metal; the bearings are usually of bronze lined with Babbitt metal, and are given large area. Crank-shafts are either of steel or of

10 H. P. VERTICAL ENGINE.—N. Y. S. S. P. Co.

hammered iron. As is customary with all well constructed engines, these engines are set up and operated in the shop long enough to exhibit all defects and to afford opportunity to make all adjustments before sending them out, and are thus made safe against those annoying delays which otherwise attend the introduction of such machines. The parts are made to gauge, and therefore interchangeable; and it is thus made easy to replace them when worn or injured, at minimum expense and with little delay. The

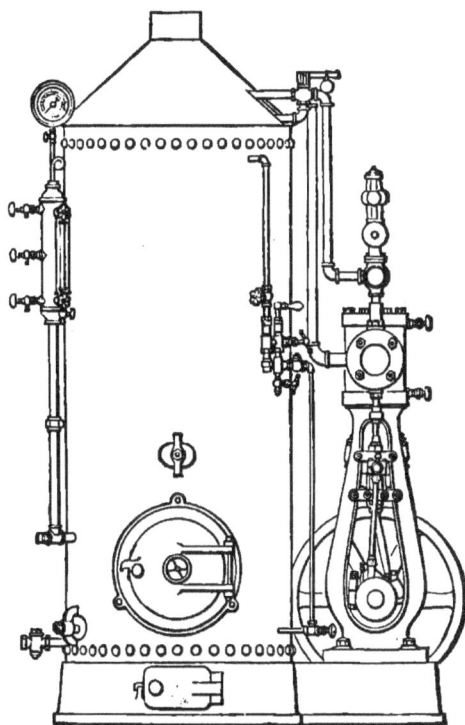

SEMI-PORTABLE ENGINE.

valves, and their seats, even, when worn, are taken out, sent to the shop, and the spare valve and seat, already fitted, takes the place of the parts removed.

Where engines are of large size, they usually have the engine room and boiler room distinct; with these small engines, however, it is found often to be desirable to place engine and boiler side by side, and even upon a common base, as is illustrated by the last of the preceding engravings. This forms what is known, frequently, as the "semi-portable" engine, to distinguish it from the "portable," which last named style is mounted on wheels.

————

THE ERICSSON AND WESTINGHOUSE ENGINES.

ALL of the engines which have been considered in the preceding articles are of one general type—that known as the "double-acting reciprocating engine." Before the time of James Watt, the only engine in extended use, even in the limited field in which the steam engine was then employed—that of pumping water from mines—was a "single-acting" engine—the Newcomen engine, which had then almost entirely superseded the so-called engine of Savery. Watt invented, first, the separate condenser, and then the double-acting engine, thus increasing the power of the machine and rendering it, at last, applicable to the turning of a crank and the driving of machinery and mill-work. In the "single-acting engine," the steam drives the piston in but one direction, and the return stroke must be made without the production of useful work. In the "double-acting engine," the steam acts upon the piston in both directions, and with practically equal effect. Thus, a more regular action is secured with a given weight of balance wheel, or the same regularity with a wheel of one-half the weight of that required for the older form of engine. This smoothness of motion is, in such work as is here considered, one of the most essential features of the best steam engine economy. At the speeds which have been now attained, however, the inertia of moving parts becomes so great that moderate variations in the impelling power become comparatively insignificant, and have no perceivable effect upon the smoothness of revolution of the crank-shaft.

The double-acting engine evidently possessed greater power than its predecessor, when of the same size, and the "efficiency of the machine" was correspondingly increased.

The very conditions which have been thus made to aid in securing regularity have, however, introduced a new difficulty: At every revolution of the engine, the crank "turns the centre" twice; and, at every passage of the centre, the direction of pressure upon the crank-pin is reversed, thus producing a shock which is proportional to the difference of pressure, the suddenness with which it is felt at the pin, and the extent of the "lost motion" between the pin and its bearings. Some lost motion must always be permitted here, to avoid danger of heating of the journal and injury to the machine. The counteracting adjustments are found to be, usually, the utilization of the inertia of the reciprocating parts, as in the Porter-Allen engine; the adoption of heavy compression, as in the several engines afterward described, and very careful adjustment of the fit of the brasses on the pin. With the skilful use of these expedients, and with the introduction of a perfection of workmanship, and of such qualities of material, as have never before been seen, the " high-speed engine " has been made successful at as high as 400, and even, in some cases, 600 revolutions per minute. The lower of these figures may be taken as that representing the maximum in standard, and usually best, practice.

But much higher speeds than these are sometimes demanded; and engines must, in the future, be built to run, regularly, steadily, and safely, at, probably, very much higher velocities. This may, ultimately, lead to radical changes in the design of the now standard forms of fast engines. Nevertheless, the limit of speed has by no means been reached, even at the higher of the above speeds, with the common type of engine. The speed of even 450 times the cube root of the length of stroke, now a common figure, and three times that given by James Watt's rule, is occasionally greatly exceeded. Captain Ericsson designed an engine, some three years ago, for the electric lighting apparatus of the Delamater Iron Works, which has now been running, every evening for two or three years, at 1,250 revolutions per minute, without giving the slightest trouble, or meeting with the most insignificant accident. The piston speed is about twice that of the average " high-speed " engine, and six times that adopted by Watt. It is probably the highest speed ever attained by a reciprocating engine doing work for which it had been designed.

The object of the inventor was to design a steam engine for the special work of driving small dynamo electric machines, and hence to secure great stability and strength, a minimum number of parts requiring lubrication, and absolute certainty that the parts retained should be, at all times, thoroughly supplied with the lubricant. The engine is therefore made a "half trunk" engine, the trunk, *F*, *F*, serving as an oil reservoir. The joint in the eccentric rod is provided with a piston moving in a cylindrical guide, *N*, which is also an oil reservoir. The cylinder, *C*, and base-plate, *B*, are in one casting, upon which is set the

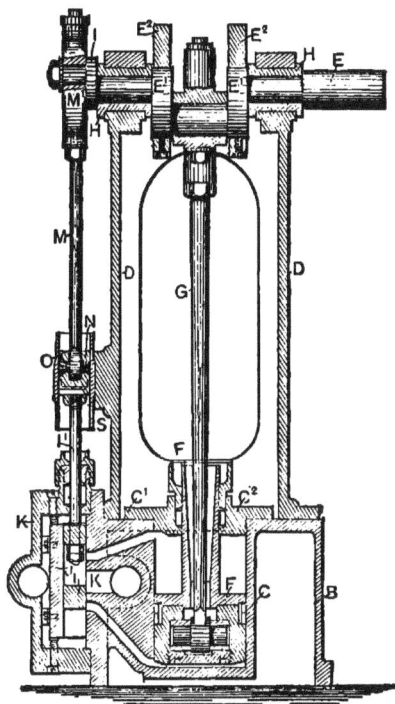

THE ERICSSON ENGINE.

hollow frame supporting the crank-shaft, *H*, *E*, and balance wheel. Every journal and rubbing part has an oil reservoir and special provision for effective lubrication. The whole engine is a model of the product of that most efficient kind of ingenuity which seeks definite ends by the most simple and directs means.   Its performance leaves nothing to be desired.

The limits to velocity of piston and speed of rotation have, from the beginning of steam engine practice, been thus gradually set farther and farther back; and one after another of the limiting conditions have been successfully met and overcome.   The earliest limit was that found in the bad workmanship and material which Watt and his contemporaries encountered, and which gave rise to heated journals at even what would now be considered very low speeds, and at very small powers.   This defect being gradually overcome, the next, and a comparatively modern, difficulty was found in wear, and the "pound," which took place when the lost motion of journals in the line of the connecting rod was taken up, at the passing of the centres.   This difficulty was met in two ways, as already repeatedly stated—by making use of the inertia of the reciprocating parts, as was done by Porter and Allen, and by heavy compression as is practiced in nearly, or quite, all of the high speed engines of to-day.   The first method can be adopted only when careful proportioning, after calculation, of the weights and velocities of the moving parts, has determined the proper weights of the compensating pieces.   The latter adjustment may be made either by calculation or by experimentally finding the compression giving smoothest running.   This effect of increasing compression can be most satisfactorily seen in the marine engine, in which, whatever the speed of the machine, and whatever the steam pressure, or however loose the journal, the link may be raised so as to gradually check the pounding at the centres, and finally to eliminate it altogether, the engine often being thus brought to work silently and smoothly at speeds far above those which, without compression, would be very troublesome, if not absolutely dangerous. This is an experiment which the writer has repeated on many engines, and almost invariably with the same satisfactory result.

Some lost motion must always be permitted at the crankpin, and these expedients are usually found to meet the case. They probably have their limits, however. There comes a time, as speeds are increased, when the weight of running parts, as calculated for strength only, becomes as great as is desirable to effect the compensation by their inertia ; there comes a time, as compression is increased, when the "cushioned" steam is carried up to boiler pressure, and this would seem the natural limit in this matter. The next device, chronologically, adopted by the engineer, is that of preventing the lift of the brass of the crank-pin and of the crosshead pin at the turning of the centres, while still leaving the freedom of fit required to give safety from heating. This last expedient is that which has led to the construction of a class of engines which are as peculiar and as typical as either of the classes which have been already described.

THE WESTINGHOUSE ENGINE

belongs to this new class, and is here taken as its representative. The change of construction characteristic of this type of engine is a return to the original "single-acting" plan of engine. This has been often proposed, and not infrequently attempted ; but the success attained has not, as a rule, been satisfactory. Two, and three, and four, cylin-

ders have been tried, in the endeavor to secure regular motion while taking steam only on one side of the piston ; very

THE WESTINGHOUSE ENGINE.

high speeds of revolution have been attained ; but the cost
of steam has been found too great, and their use has not
become general. The Westinghouse engine has proved it-

self to possess the elements of commercial success, and is, therefore, to be taken as illustrating what can be done in this direction, by good designing and good business management.

It is evident that, if steam pressure comes upon but one side of the piston, the engine can pass its centre without the brass lifting clear of the pin, and thus may be driven up to any speed without liability of injurious pounding. For enormously high speeds, as the engineer of to-day looks upon them, this is evidently the type of engine to be looked to for smooth and successful working. The illustrations show how, in the Westinghouse engine, this end is reached. The engine has two cylinders, A, A, fitted with single-acting pistons, D, D, forming trunks filling the bore of the cylinder, giving a long steam-tight bearing, and taking the connecting-rod pin, A, B, at a point at which no tendency to rock the piston can be produced. The top of the piston is cored out to prevent transfer of heat from the working to the non-working end. The rods, F, F, take hold of the crank-pins within an enclosed chamber, C, forming part of the engine frame, E, C. This frame and bedplate also acts as a reservoir for oil lubricating the journals and pistons, which oil floats on water and is dashed up over the moving parts so enclosed, at every revolution of the engine. No other attention is required than to keep a supply of oil in the chamber, by filling as loss occurs by leakage. In fact, the whole engine is thus shut in by its frame, and its working parts are invisible, while working—an arrangement at once a means of security and convenience.

The valve adopted in the Westinghouse engine is a piston valve of the class already described, but having some peculiarities specially adapting it to its use in this engine. Its guide, *J*, is a piston traversing a cylinder separating the exhaust space from the chamber below. This one valve, *V*, distributes steam to both cylinders, the two cranks being set directly opposite each other. This adjustment of the cranks also gives a perfect balance of reciprocating parts, and secures smoothness of movement of the whole machine, whatever speed may be adopted ; and exceptional speeds of 1,000 revolutions, or more, per minute are reached without observable vibration.

The governor, *I*, and its action, are precisely like the same parts in the engines described in several of the earlier articles. It actuates the eccentric, and determines the point of cut-off by varying the throw of the valve, while retaining the lead. The governor is usually so adjusted that it will not come into play until the engine falls 1 per cent. below, or rises 1 per cent. above, the normal speed ; its full traverse is effected, also, within this range, the intention being that the speed shall never vary more than 1 per cent. from that fixed as its proper velocity. The range of expansion is from o to about 5-8 stroke.

One of the dangers to which fast running engines are peculiarly exposed is that of injury by the entrapping of water in the cylinder, and the plunging of the piston against the mass of incompressible fluid which then fills the clearance spaces. In this engine, in addition to the relief-cocks, or valves, which are always fitted to such engines, a safeguard is introduced in the form of what engineers are accustomed to call the "breaking-piece," a part which is made purposely weaker than other portions of the machine, exposed to a common danger, so that this piece may go

when danger arises. This piece is always one the replacement of which will give little trouble, and make but little
expense. In the Westinghouse engine, such a breaking-
piece is made to form a part of the cylinder head. This

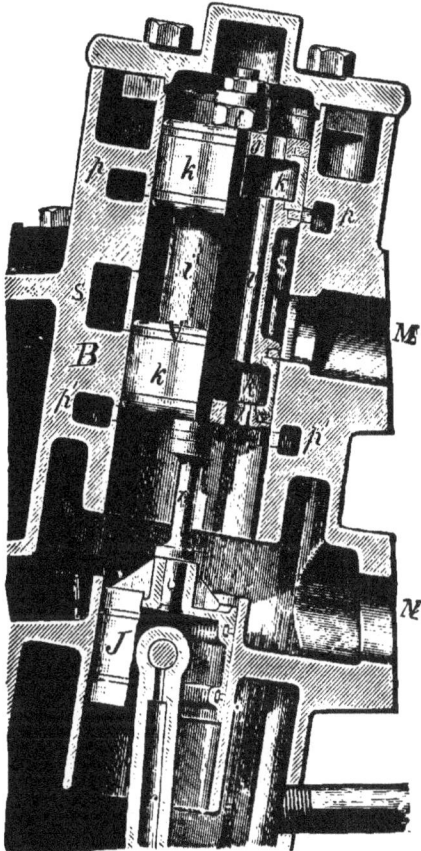

The Westinghouse Engine.—Cross Section Through Valve.

may be knocked out without injury occurring to any important, or costly, part of the structure.*

* The writer planned an engine, about the year 1860 in which the whole
cylinder-head was made a safety valve which could lift and discharge the water
into the chamber behind it, the cover of the latter being bolted on, while the
cylinder-head was only held in place, against a faced joint, by steam pressure.

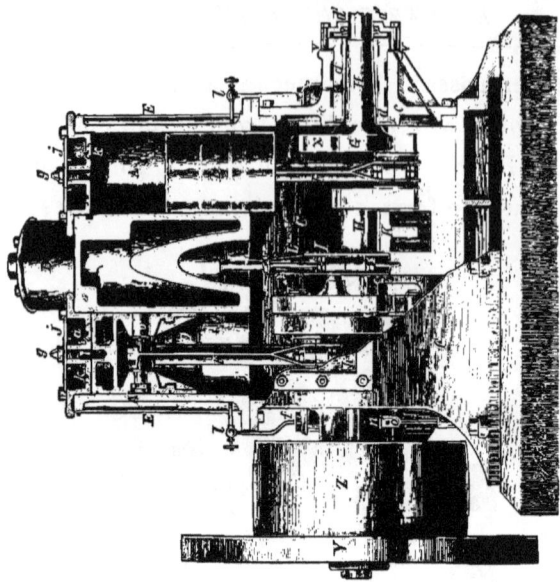

THE WESTINGHOUSE ENGINE.—SECTION THROUGH SHAFT

.

This breaking-piece is intended to yield at a safe pressure—200 lbs. per square inch— and thus save the engine. The workmanship on these engines, so far as the writer has been able to examine it, is excellent ; and the material of the best.   These are, however, as has been stated, absolutely essential features of every good high-speed engine.   The engines are, when finished, set up in the shop and tested up to their rated power, before sending them out ; and it is thus made certain that they are in good order and in correct adjustment.   The ingenious and novel methods of securing certainty of lubrication, in this engine, the constant direction of the actions tending to produce heavy strains, the small number of parts subject to wear and to breakage, the remarkable success met with in the attempt to reduce the labor of attendance and cost of maintenance, and all other costs causing reduction of commercial efficiency ; the compactness, solidity, steadiness, safety at maximum speeds, and general effectiveness of this engine, are such as to make it one of the most interesting examples of the steam engine of to-day that has yet attracted the attention of the engineer.

The economy of the later style of this engine—that fitted with automatic expansion gear, as here described— is probably about the same as that of other small engines of its own class ; not, as a matter of course, equal to the economy of large engines of the four-valve type, but great as compared with the class of small engines to which the manufacturer has usually been compelled to resort up to the present time, when demanding but little power.   The loss by "friction of engine" is somewhat greater in this form than in the more familiar type of engine.   The

peculiar advantages possessed by the engine in this direction are its high piston and rotative speed, and the
extent to which compression is carried. One of these
engines has been driven experimentally up to 2,700 revolutions per minute without any observable ill effect.
Their speeds are probably safely made double that of the
average "high-speed engine" with which we are now
becoming familiar. Compression, as an element in economy of engine, has already been considered at some
length. It was shown by the writer, ten years ago, that
progress in the direction of improvement of the steam
engine has always been retarded by the difficulty of preventing serious losses by cylinder condensation, and that
this is the essential element of preventible waste ; it was
also suggested by him, several years since, that probably
the best means of controlling the speed of engine is by the
introduction of high compression and its variation by
the governor, increasing compression with increasing expansion, and the reverse, and thus, by utilizing the heat of
compression, checking cylinder condensation as its increase
is caused by extending the expansion period ; and it was
pointed out that "the best among existing forms of valve
gear should, if judged by from the standpoint here taken,
be that which, combining a variable expansion with a variable
compression, is also capable of prompt and exact adjustment by a sensitive and efficient governor."* This sug-
gestion, as has been seen, is fully met by all the later
designers of engines of the high-speed class. The engine

---

* Expansion of Steam, etc., Trans. Am. Soc. Mech. Engrs., 1881; Jour. Fran.
Inst., Oct. 1881.

above described illustrates well this use of compression; the compression is adjustable by the governor, and may thus given be that ratio which is best adapted to the case.

Mr. Harris Tabor, in a paper read before the American Society of Mechanical Engineers, following the idea just presented, says of compression: "It is to the proper control of compression that we must now look for further advance in steam economy." It has been seen that this is one of the directions of present advance.

RETROSPECT.—We have now made a tolerably complete survey of the whole modern field of steam engineering as far as it is covered by stationary engine practice, and have seen a very steady progress from the best types of a generation ago to the most representative examples of the most modern forms. It is seen that the direction of change is still that which, as has been often pointed out by the writer, has been observed from the days of James Watt. The principal points found worthy of notice have been the increase in economy and general efficiency by a tentative and empirical, but none the less steady and uninterrupted, method of advance. The pressures of steam have been slowly, but constantly, rising; speeds of piston, and of rotation, have been as constantly increasing; the effectiveness of the governor has been made greater and greater; the ratio of expansion at maximum efficiency has been very slowly increased, by the gradual reduction of cylinder condensation; commercial considerations have been brought definitely into view; the efficiency of engine has been improved by reduction of size, weight, and friction of engine; and thus we have been able to see a gradual change of type of engine effected, the engineer modifying his designs to meet the demands of the time, until we have insensibly, and almost without suspecting that progress has been going on, passed across a new line and entered upon an epoch, in steam engine construction, as marked in its period and as well defined, as to its beginning, as was that which, at the middle of the century, was distinguished by the introduction of the inventions of Sickles, Corliss, and Greene.